Advance Praise for

Power from the **Wind**

When you need practical advice from a warm, smart
and informed human being, Dan Chiras is the one to turn to.

— Bruce King, PE Director, Ecological Building Network;
and author, *Buildings of Earth and Straw* and *Making Better Concrete*

Dan Chiras has done as much as anyone in America to
promote and popularize the use of renewable energy.

— Stephen Morris, publisher and editor,
Green Living: A Practical Journal for the Environment; and editor,
The New Village Green: Living Light, Living Local, Living Large

Dan Chiras is one of the most authoritative writers
in the field of renewable energy. His multiple books create
a comprehensive library for homeowners looking to live
a lifestyle in harmony with their values.

— David Johnston, What's Working: Visionary Solutions
for Green Building; and author, *Green Remodeling:
Changing the World One Room at a Time*

power
from the
wind

DAN CHIRAS
with Mick Sagrillo
and Ian Woofenden

Technical Advisors
Robert Aram, PE
Jim Green, PE

NEW SOCIETY PUBLISHERS

CATALOGING IN PUBLICATION DATA

A catalog record for this publication is available from the National Library of Canada.

Cover design by Diane McIntosh.
Cover Image: iStock

Printed in Canada.
First printing February 2009.

Paperback ISBN: 978-0-86571-620-9

Inquiries regarding requests to reprint all or part of *Power from the Wind* should
be addressed to New Society Publishers at the address below.

To order directly from the publishers, please call toll-free (North America)
1-800-567-6772, or order online at www.newsociety.com

Any other inquiries can be directed by mail to:

New Society Publishers
P.O. Box 189, Gabriola Island, BC V0R 1X0, Canada
(250) 247-9737

New Society Publishers' mission is to publish books that contribute in fundamental ways to building an
ecologically sustainable and just society, and to do so with the least possible impact on the environment,
in a manner that models this vision. We are committed to doing this not just through education, but
through action. This book is one step toward ending global deforestation and climate change. It is printed
on Forest Stewardship Council-certified acid-free paper that is **100% post-consumer recycled** (100% old
growth forest-free), processed chlorine free, and printed with vegetable-based, low-VOC inks, with covers produced using FSC-certified stock. Additionally, New Society purchases carbon offsets based on an
annual audit, operating with a carbon-neutral footprint. For further information, or to browse our full list
of books and purchase securely, visit our website at: **www.newsociety.com**

NEW SOCIETY PUBLISHERS

Dedication

To my son Forrest, a kind, gentle, loving and talented soul
who graces us all with his presence.

Books for Wiser Living
recommended by Mother Earth News

Today, more than ever before, our society is seeking ways to live more conscientiously. To help bring you the very best inspiration and information about greener, more sustainable lifestyles, *Mother Earth News* is recommending select New Society Publishers' books to its readers. For more than 30 years, *Mother Earth* has been North America's "Original Guide to Living Wisely," creating books and magazines for people with a passion for self-reliance and a desire to live in harmony with nature. Across the countryside and in our cities, New Society Publishers and *Mother Earth* are leading the way to a wiser, more sustainable world.

Contents

Preface

My work on this book started in the summer of 2006. I'd just published *The Homeowner's Guide to Renewable Energy*, a book that helps readers understand their options for tapping into renewable energy, especially solar and wind.

E-mails and phone calls started arriving almost immediately, with technical questions about solar electricity and wind. Readers assumed I was an expert in every renewable energy technology discussed in the book, when, in fact, my area of greatest expertise at the time was in passive solar heating and cooling. I'd published a book a few years earlier on that topic entitled *The Solar Home: Passive Heating and Cooling*. Although I'd installed my own wind system and had solar hot water collectors installed on a previous home, and lived off-grid on a solar electric system in my super-efficient passive solar home in the foothills of the Rockies for many years, I certainly wasn't an expert on wind and other types of renewable energy systems other than passive solar.

In the summer of 2006, a few months after *The Homeowner's Guide to Renewable Energy* was published, I was visiting a client in northern Michigan, helping her design a super-efficient passive solar facility for a small organic farm and ecological learning center. She asked if I'd help her design a wind energy system to produce electricity to pump water to irrigate her crops. I said yes reflexively. What could be so hard about that?

When I arrived home, I began to read everything I could put my hands on about wind to expand my knowledge. For reasons that will probably always be a mystery to me, I became obsessed with this amazing technology and all its technical intricacies. Whatever I learned, I had to know more.

As my knowledge grew, I felt the urge to put it all down on paper, to write a book for those who want to learn about wind but don't want to wade through a hundred books and articles on the subject. As my knowledge grew, however, so did my awareness that this was a subject that required an extremely high level of expertise. I quickly came to realize that I needed a really smart, experienced and knowledgeable expert in small wind energy to help me — to provide advice and guidance, correct inadvertent mistakes, and provide additional information. While attending the Midwest Renewable Energy Association's annual energy fair, I bravely asked Mick Sagrillo, a guy who fit the bill precisely, if he'd help. Mick's the guru of small wind. He has been in the small wind business longer than just about anyone. Mick said that despite his hectic schedule, he'd assist me.

Mick Sagrillo has been in the small wind industry since 1981. He teaches workshops on wind energy through the Midwest Renewable Energy Association and Solar Energy International. He writes a column for *Windletter*, the American Wind Energy Association's newsletter, and *Solar Today*, and has published numerous articles in *Home Power* magazine. He consults on Wisconsin's Focus on Energy's small wind program and served as the lead technical advisor to me on this project, offering invaluable information and advice throughout.

A year or so later, I approached New Society Publishers with the idea of publishing a series of books on renewable energy technologies for the home and office. They liked the idea and we were off and running.

After bringing Mick on, I stepped up my education. I signed up for numerous workshops on wind energy through various entities like the Midwest Renewable Energy Association and the American Solar Energy Society. Most important to my learning were hands-on workshops on the installation of wind turbines, wind turbine design, and wind site assessment. In late 2007, I became a certified wind site assessor — at the time only one of twelve in the nation.

That year, I also recruited another wind expert, *Home Power's* Ian Woofenden. Like Mick, Ian teaches installation workshops and writes articles for *Home Power* magazine on wind energy. Ian supplies a good portion of his electricity from the wind and solar, too, and has been living off grid for many years.

In the summer of 2007, Jim Green, the National Renewable Energy Laboratory's small wind expert asked if he could help, too, after I attended a workshop of his on small wind energy, which he co-taught with Abundant Renewable Energy's Robert Preus at Solar 2007 in Cleveland. Jim volunteered to read the manuscript and provide his insights and to help ensure the technical accuracy of the book.

A few months later, I recruited another expert, Robert Aram, an electrical engineer, who I met and worked with in several wind energy workshops. Bob impressed me with his vast knowledge and extraordinary ability to explain things clearly, as well as his no-nonsense insistence on correct math, physics, and engineering terms. He said that he'd be happy to read the book to help ensure technical accuracy. I hired him to give me a hand, to be sure that this book, while written for the nonengineer, was correct in every aspect.

Mick, Ian, Jim and Bob offered extremely valuable comments that helped me produce what I hope will be the most accurate, up-to-date and readable resource on small wind energy. Without them, this task would have been impossible. I am deeply indebted to all of them for their tactful and honest comments and commitment to accuracy and am extremely honored to have worked with such amazingly dedicated and knowledgeable people. A world of thanks to them and to all the others whose work I relied on when researching and writing this book.

A world of thanks also to my dear friends at New Society Publishers, Chris and Judith Plant, who have, over the years, been an absolute delight to work for. Thanks for taking this project on and for believing in me, and thanks for their unwavering dedication to cre-ating a sustainable society. Thanks to the staff at New Society as well, Ingrid who shepherded this book through the process, Greg Green and MJ Jessen did a smashing job of designing and laying out the book, and to my copyeditor, Murray Reiss, for helping make this a better book. I'd also like to thank a new member of my team, Dr. Anil Rao, a professor of biology, who illustrated this book for me. He's been a pleasure to work with and a valuable part of this book's success.

Of course, I'd also like to thank my family. First a world of thanks to my partner Linda, who nursed me back from a bad case of bacterial pneumonia during the final stages of this book's writing and who has listened intently to my many discussions of wind turbines and wind site assessments. Thanks, too, to my sons, whose lives continue to grace mine.

Dan Chiras
Evergreen, Colorado
June 2008

CHAPTER 1

INTRODUCTION TO
SMALL-SCALE WIND ENERGY

Humans have harvested energy from the wind for centuries. Prior to the advent of steam-powered ships, for example, Phoenicians, Europeans and others relied on the wind to propel magnificent sailing vessels across a largely uncharted planet. Soon ships became an important mode of transport for raw materials and finished products to and from Europe.

Our predecessors also used wind to assist in food production and to manufacture goods. The windmills of Europe, for example, which were in place 800 to 900 years ago, were used to grind grain into flour to feed Europe's masses. The Dutch used wind to pump water from coastal wetlands, so they could be converted to farmland to grow food.

Wind energy has a long history in North America, too, stretching into the late 1800s. During this period, windmills on tens of thousands of farms in the Great Plains of North America pumped water for livestock, garden sand humans. Without them, many farmers would not have been able to provide sufficient water for their cattle and sheep — or themselves.

Although history books make little mention of it, in the 1920s through the early 1950s many Plains farmers also installed small wind turbines to generate electricity. These electric-generating wind turbines made life on the Great Plains more bearable. Home-grown electricity was used to power lights and a handful of modern conveniences, among them electric

Windmill vs. Wind Turbine

A windmill is a device that drives a mechanical load such as a water pump. A wind turbine or wind generator drives an electrical generator.

toasters, washing machines and radios, all ordered from the Sears catalog. The radio was highly coveted as a way of keeping in touch with the world. Purchasers of a radio were given a discount on their wind generator.

Both wind-electric generators and water-pumping windmills were extremely popular among farmers and ranchers.

In the 1890s, more than 100 manufacturers produced water-pumping windmills in the

Fig. 1.1: *Water pumping windmills like this one, photographed by Dan on a commercial wind farm in southeastern Colorado, were once common through the West and Midwest. The technology is so good that it hasn't changed in 100 years.*

United States, notes small wind expert Jim Green of the National Renewable Energy Laboratory. According to the US Department of Energy's National Renewable Energy Laboratory, over 8 million mechanical windmills (water pumpers) were installed in rural America, beginning in the 1860s (Figure 1.1). Many of these water-pumping windmills have been restored and are still operating today, providing many years of reliable service with minimal maintenance.

Wind energy was not only vital to farmers, it was extremely important to railroads in the wild West — a fact largely ignored by historians. Windmills were often used to fill water tanks along tracks to supply the steam engines of early locomotives.

Unfortunately, the use of water-pumping windmills and wind-powered electric generators began to decline in the United States in the late 1930s. The demise of these technologies was due in large part to America's ambitious Rural Electrification Program.

This program, which began in 1937, was designed to provide electricity to rural America. As electric service became available, wind electric generators were mothballed. In fact, local power companies required farmers to dismantle their wind generators as a condition for providing service via the ever-growing electrical grid. The electrical grid, typically referred to as the grid, is the extensive network of electrical transmission lines that criss-cross our nation, delivering electricity generated by centralized power plants to cities, towns and rural customers. A key advantage of the grid was its ability to provide virtually unlimited amounts of electricity to those who had the wherewithal to pay for it. The grid also made it possible to power large motor loads, something that wind/battery systems were unable to do.

Although farmers' lives improved as a result of rural electrification, once-profitable manufacturers of wind-electric generators were driven out of business by the early 1950s. In the mid 1970s, however, wind energy made a resurgence as a result of intense interest in energy self-sufficiency in the United States and elsewhere. This new-found interest in self-reliance was stimulated principally by back-to-back oil crises in the 1970s that resulted in skyrocketing oil prices and a period of crippling inflation in the United States. Extremely generous federal incentives for small wind turbines (a 40 percent US federal tax credit), equally charitable incentives from state governments, and changes in US law that required utilities to buy excess electricity from small renewable energy generators helped spark the comeback. From 1978 to 1985, 4,500 small utility-connected residential wind machines were installed, according to Mike Bergey, whose company Bergey Windpower manufacturers small wind turbines. In addition, approximately 1,000 wind machines were installed in remote locations not connected to the electrical grid.

In short order, however, wind energy's resurgence died, falling victim to economic forces beyond its control. Energy efficiency measures in the United States and new, more reliable sources of oil from Great Britain, Russia and other countries, drove the price of energy downward. These factors, combined with the end of federal and state renewable energy tax incentives and a dramatic shift in the political climate away from renewable energy in the early 1980s, resulted in a precipitous decline in America's concern for energy independence. As a result of these changes, most of the fledgling wind manufacturers went out of business. In fact, six years after the end of the tax credits, virtually all of the 80 or so wind generator companies doing business in the United States disappeared, according to Mick Sagrillo, the small wind energy expert who served as a technical advisor for my writing of this book.

In the 1990s, after nearly two decades of quiescence, commercial and residential wind

Fig. 1.2: *Mick Sagrillo, perched on a tower in this photo, has been in the small-scale wind industry since 1981.*

Rated Power in Watts or Kilowatts

Wind turbines are commonly described in terms of rated power, also known as rated output or rated capacity. Rated power is the instantaneous output of the turbine (measured in watts) at a certain (rated) wind speed and at a standard temperature and altitude. Small wind turbines, the subject of this book, have a rated power of 1,000 to 100,000 watts. One thousand watts is one kilowatt (kW). Small wind falls in the range of one kilowatt to 100 kilowatts. Large wind turbines include all of those turbines over 100 kilowatts. Most larger turbines, however, are one megawatt and larger machines. A megawatt is a million watts or 1,000 kilowatts.

While rated power is commonly used when describing wind turbines, it is one of the least useful and most misleading of all parameters by which to judge a wind generator's performance, for reasons we'll make clear in Chapter 5.

As you study wind energy and other energy systems, you'll commonly hear experts talk about the "capacity" of a wind farm or a wind turbine. They'll talk about a 20-megawatt wind farm. The capacity of a wind turbine is its rated output. If you installed ten 20-kilowatt wind turbines, you would have installed 200 kilowatts of wind generation capacity — that is, 200 kilowatts of electric- producing *potential*. However, these wind machines do not produce this amount of electricity all of the time, only when they're running at their rated wind speed. At their rated wind speed, ten 20-kilowatt wind turbines should produce 200,000 watts of electricity.

energy made another comeback. This most recent rise in the popularity of wind and other renewable energy resources was spurred by deep concern over rising energy prices. However, several other factors have played a significant role in wind's latest resurgence: (1) profound concern for the decline in world oil production, (2) the sharp increase in price of natural gas and declining supplies in the United States and Canada, and (3) concern for global climate change and its costly impacts.

Because of these factors, many proponents of renewable energy believe that this time around, wind energy is here to stay. It's hard to argue with them. Evidence is everywhere. Much to the delight of renewable energy advocates, large commercially operated wind farms are popping up on land and in the sea in numerous countries, most notably the United States, Germany, Spain and Denmark. Commercial wind farms crank out huge amounts of electricity and promise to significantly change the way modern society meets its energy needs.

Today, wind-generated electricity is the fastest growing source of energy in the world (Figure 1.3). In 2006, over 15,000 megawatts of wind capacity were added (capacity indicates the full production of a power source under optimal conditions). In 2007, 20,000 megawatts of capacity was added, according to the Global Wind Energy Council. These additions are more than three times the wind capacity added in 2000 and twelve times the capacity added in 1995.

Although commercial wind farms are responsible for most of the growth in the wind industry, smaller residential-scale wind machines are also emerging in rural parts of America and other countries, supplying electricity to homes, small businesses, farms, ranches and schools (Figure 1.4). Even a few large businesses have installed small wind machines (under 100 kilowatts) to power their facilities. Most of the small-scale wind turbines "feed" the excess electricity they produce back onto the electrical grid.

A handful of wind pioneers have also explored ways to capture the energy of the wind to heat homes, although they have met with very limited success in the marketplace. In addition, many sailboats are equipped with very small wind machines under 1,000 watts — typically referred to as microturbines — to power lights, fans and refrigerators (Figure 1.5). Ranchers and farmers sometimes use wind turbines to supply power to electric fences, stock watering tanks and remote lighting — that is, small dedicated loads to which it is not cost effective to run a power line. I've seen small wind turbines used to power park facilities in remote locations in Alaska.

Wind energy is being tapped to power remote villages in less developed countries, where the cost of stringing power lines from centralized power plants is prohibitive. Wind energy has even found a home in remote sites in some developed countries. In France, for instance, the government paid to install wind

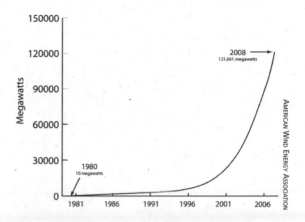

Fig. 1.3 (above): *Global Wind Energy Capacity. This graph shows the installed global capacity (in megawatts) of commercial wind turbines. In 2007, the global capacity was the equivalent of 188 500-megawatt coal-fired power plants.*

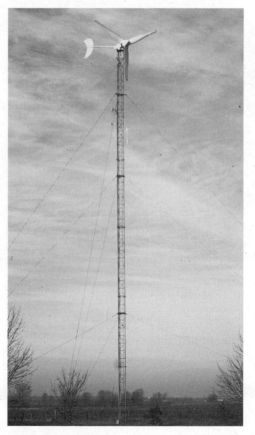

Fig. 1.4 (left): *Small Wind Turbine on Tower. This ARE442 wind turbine installed at Mick's house during a workshop is mounted on a guyed lattice tower. Maintenance is performed by climbing the tower.*

Fig. 1.5: *Wind Turbine on Sailboat. (a) Microturbines, such as the one shown here by Marlec, are frequently used on sailboats to charge batteries that supply electricity for loads such as radios, lights, televisions and refrigerators. (b) Microturbines for marine use are designed to withstand the harsh environment. This one is made by Aerogen.*

turbines and solar electric systems on farms at the base of the Pyrenees, rather than running electric lines to these remote operations. Even nomads in Mongolia tote tiny wind machines to provide electricity to their yurts. When they move on every few weeks in search of new pasture for their livestock, their wind machines are packed up and transported on the backs of pack animals.

Wind clearly has a long history of service to humankind, and is on the rise. Proponents say it could become a major source of electricity in years to come.

World Wind Energy Resources

Although wind energy's popularity is at an all-time high, and continues to grow yearly, what is its potential? Can wind become a major source of energy in the future?

Wind is a ubiquitous resource. Although not evenly distributed throughout the world, significant resources are found on every continent. Globally, wind resources are phenomenal. Tapping into the world's windiest locations could theoretically provide 13 times more electricity than is currently produced worldwide, according to the Worldwatch Institute, a Washington, DC-based nonprofit organization that's played a huge role in creating a sustainable future.

In North America, wind is abundant much of the year in the Great Plains and in many northern states. It is also a year-round source of energy along the Pacific and Atlantic Oceans and the shores of the Great Lakes. Tapping into the windiest locations in the United States, for example, in North and South Dakota — or North Dakota and Texas — could produce enough electricity to supply *all* of the nation's electrical needs. Proponents of wind energy, like the Worldwatch Institute, estimate that wind energy could provide 20 to 30 percent of the electricity consumed in many countries. Others believe that wind could provide an even larger percentage.

Could wind provide 100 percent of the world's electrical energy needs?

Yes, it could, theoretically … if we are creative in storing electricity and also transferring it throughout countries via the grid.

Will it?

Probably not.

Other sources of renewable and nonrenewable energy will also play a role in meeting our energy needs.

In the future, however, wind will very likely play a huge role in many parts of the world. Commercial and, to a lesser extent, residential wind turbines will produce enormous amounts of electricity for homes, businesses, farms and ranches. Commercial solar electric facilities and solar-electric systems on homes and businesses could also produce a significant amount of electricity. Solar thermal electric systems, typically referred to as concentrated solar power, operated by large utilities could add to the mix.

The potential of the sun, like that of the wind, is nothing short of phenomenal. It's estimated, for instance, that the sunlight striking an area the size of the state of Connecticut could meet all of the United States' (inefficient

> ## Can Wind Do it All?
>
> As Ian Woofenden, wind energy expert, author, and technical advisor to this book points out, "The 'Can wind do it all?' question is a bit of a red herring. "Wind is one piece in the puzzle; nothing is the whole answer. What we need is movement in the right direction. If we don't get all of our energy from renewable sources, the world will not collapse. If we don't start moving in that direction, it might."

and wasteful) electrical demand. Although no one is proposing the construction of such large solar-electric arrays, solar-electric modules on homes, office buildings, schools and commercial solar-electric facilities in the best locations could provide an enormous amount of electricity, supplementing wind energy production (Figure 1.6).

Geothermal and biomass resources could contribute their share as well. Biomass resources refer to plant matter such as wood chips that can be burned directly to produce heat to generate steam to make electricity. Plant matter such as corn can also be converted into gaseous or liquid fuels that can be burned to create electricity. Animal wastes can also be used to generate methane, the main component of natural gas.

Hydropower will continue to do its part in the future, and lest we forget, conventional fuels such as oil, natural gas, coal (burned as cleanly as possible), and nuclear energy will also be part of the mix for many years to come.

Despite what some critics contend, renewable energy, including wind, is here to stay and will likely contribute significantly to our energy future. It has to for the simple reason that fossil fuels are limited. Oil could be economically depleted within 30 to 50 years. Production rates worldwide are on the decline now. Natural gas production could also peak in the not-too-distant future. The sun, however, which powers solar energy systems and creates winds that can be tapped by wind turbines, is going to be around for at least 5,000,000,000 years.

The Pros and Cons of Wind Energy

Wind is a seemingly ideal fuel source that could ease many of the world's most pressing problems. Like all energy sources, wind power has its advantages and disadvantages. Let's look at its downsides first.

Disadvantages of Wind Energy

As you read the downsides of wind energy, you'll discover that many of them pertain to large commercial wind projects. These concerns, in turn, trickle down unfairly to small

Fig. 1.6: *Solar Array. In a renewable energy economy, large-scale solar electric installations, like this one, will supplement electricity produced by other renewable resources, including wind, hydropower and biomass, as well as conventional fuel sources.*

DAVID AMSTER

wind. You'll also see that, while there are valid problems with wind energy, some are perception problems — "problems" that result from misconceptions and deception on the part of opponents. We'll be sure to point these out as we proceed. We'll also counter unfair criticism so you receive a balanced view.

Variability and Reliability of the Wind

Perhaps the most significant "problem" with wind is that the wind does not blow 100 percent of the time in most locations. Like solar energy, wind is a variable resource. A wind turbine may operate for four days in a row, then sit idle for the next two days. In most locations, winds are typically strongest in the fall, winter and early spring, but die down during the summer months.

Wind even varies during the course of a day. Winds may blow in the morning, then die down for a few hours, only to pick up later in the afternoon and blow throughout the night.

Even though wind is a variable resource, it is not unreliable. Just like solar energy, you can count on a certain amount of wind each year. With smart planning and careful design, you can design a wind system to meet some or all of your electrical needs.

Wind's variable nature can be managed to our benefit by installing batteries to store surplus electricity in off-grid systems. The stored electricity can power a home or office when the winds fail to blow — or when demand exceeds the output of the turbine.

Surplus electricity can also be stored on the electrical grid in many systems. That is, when a wind-electric system is producing more power than a home or business is using, the excess can be fed onto the grid. In times of shortfall, electricity is drawn from the grid. (The grid serves as an unlimited battery bank to store excess electricity.) A grid-connected wind system can be designed to meet a small percentage of your electrical needs or all of them. (We'll discuss off-grid and grid-connected systems in more detail in Chapter 3.)

Wind's variable nature can also be offset by coupling small-wind systems with other renewable energy sources, for example, solar-electric systems or micro hydro systems. These are referred to as hybrid systems. Solar-electric systems or photovoltaic (PV) systems generate electricity when sunlight strikes solar cells in solar modules. Micro hydro systems tap the energy of flowing water in streams or rivers near homes and businesses. They convert this energy into electricity. Hybrid systems can be sized to provide a year-round supply of electricity. As you shall see in Chapter 3, residential wind-generated electricity can also be supplemented by small gas or diesel generators.

Wind's presumed unreliability even comes into question when renewable energy experts compare their systems — for example, when wind is compared to solar electricity. Some experts mistakenly view solar electricity as a more reliable resource than wind. However, the wind is much more predictable than you'd think. To understand what we mean by this,

> "As a power source, wind energy is less predictable than solar energy on a day to day basis, but it is also typically available for more hours in a given day."
>
> — Mike Bergey, Bergey Windpower

let's look at the capacity factor, a measurement used to compare different electrical generating technologies.

Capacity factor is the ratio of the actual output of a power plant over some period to what its output would have been had it operated at its rated power for the same period. For example, let's suppose you live in an area with four peak hours of sunlight for PV production per day. In this location, the capacity factor for PV would be 4 hours per day divided by 24 hours per day or about 17 percent.

In the lower 48 states, the capacity factor from most fixed PV systems ranges from 8 to 25 percent, depending on the location. According to Mick, the capacity factor of small wind systems ranges from 10 to 28 percent. So, PV and wind systems are fairly similar.

The capacity factor of wind is so high because wind turbines can work day or night — in sunny weather and cloudy weather. What is more, because wind and sunlight are often available at different times, the two technologies can complement each other extremely well. Hybrid systems increase the electrical energy produced at a site; they also reduce the hourly, daily and seasonal variation in output.

Bird and Bat Mortality

Another perceived problem that frequently arises in debates over wind energy is bird and bat mortality. Unfortunately, this issue has been blown way out of proportion for both small and large wind turbines. Although a bird may occasionally perish in the spinning blades of a residential wind machine, this is an extremely rare occurrence. Ian is aware of only one instance of a bird kill, when a hawk flew into a small wind turbine. "Because of their relatively smaller blades and short tower heights, home-sized wind machines are considered too small and too dispersed to present a threat to birds," notes Mick in his article, "Wind Turbines and Birds" published by Focus on Energy, Wisconsin's Renewable Energy Program.

The only documented bird mortality of any significance occurs at large commercial-scale wind turbines — but even then, the number of deaths is extremely small. In our view, the argument that wind energy development should be halted because of bird kills is ill-informed, or sometimes a dishonest ploy by individuals and organizations that oppose wind energy development. If citizens and governments were serious about bird kills, we'd ban the truly lethal forces discussed in the accompanying box: domestic cats, utility transmission towers, cars, pesticides and windows. We'd even prohibit farming, which destroys bird habitat and poisons birds with pesticides.

Studies also show that while bats are killed by large commercial wind turbines in certain locations, such occurrences are rare. According to researchers, large wind turbines in certain locations kill, on average, 2.45 to 3.21 bats per year.[1]

While bat deaths, like bird deaths, are regrettable, there's no indication that bat populations in the vicinity of large or small wind turbines are in any way threatened by them. Other factors play a much larger role in bat mortality, including pesticides, habitat destruction,

Bird Kills from Commercial Wind Farms: Fact or Fiction?

While commercial wind machines do kill a small number of birds, scientific studies show that the problem has been grossly exaggerated. These studies indicate that bird kills from large commercial wind turbines pale in comparison to deaths from several common sources, among them domestic cats, electric transmission lines, windows, pesticides, motor vehicles and communication towers (Table 1-1). Worldwide, hundreds of millions of birds — perhaps even billions — are killed each year by these sources. Commercial wind turbines, on the other hand, kill a miniscule number of the birds. So why has wind gotten such a bad reputation?

Table 1.1
Estimated Annual Bird Death in the United States by Source

Activity/Source of Bird Mortality	Estimated Annual Mortality
Killed by cats	270 million or more
Collisions with and electrocution by electrical transmission wires	130 to 170 million
Collisions with windows	100 to 900 million
Poisoning by pesticides	67 million
Collisions with motor vehicles	60 million
Collisions with communications towers	40 to 50 million

AMERICAN WIND ENERGY ASSOCIATION

lighthouses, communication towers, power lines, fences and human disturbance during hibernation. For more on this topic, check out Mick's article "Bats and Wind Turbines" published in the February 2003 issue of AWEA's *Windletter*.

Wind machines got a bad rap from one of America's oldest and largest wind farms: the Altamont Pass Wind Resource area in California. Located just east of San Francisco, Altamont Pass is home to a mind-boggling 7,000 wind turbines. It is also the habitat of numerous raptors. Soon after the wind turbines were erected, the birds began to perch on the wind towers in search of abundant prey (ground squirrels and other rodents) that live year-round in the grasses at the base of the towers. Some raptors died as they flew into the blades of the turbines toward prey on the ground.

A two-year study of bird kills in the region revealed only 182 dead birds in that time. While any raptor death is of concern to those of us who cherish wildlife, the death rates at Altamont from wind machines are insignificant compared to those from other factors.

Cats are probably the most lethal force that birds encounter. According to one study, a feral cat kills as many birds in one week as a large commercial wind turbine does in one to two years. Declawing a cat doesn't seem to help much. According to one researcher, the majority of cats (83 percent) kill birds, even declawed and well-fed cats prey on wild birds. Neutering or spaying a cat does not seem to cut down on hunting, either. With more than 64 million cats in America alone, what's the total loss?

No one knows for sure, but if the situation in Wisconsin is indicative of the national toll, America's bird population is being decimated by our furry feline companions. In Wisconsin alone, researchers estimate that cats kill approximately 39 million birds per year. Nationwide, the number is estimated to be around 270 million, and is very likely much higher. "Even if wind were used to generate 100 percent of US electricity needs, at the current rate of bird kills, wind would account for only one of every 250 human-related bird deaths," notes the AWEA.

Another 130 to 174 million birds die each year as a result of collisions with or electrocution by electrical transmission lines that crisscross the nation. Many victims are raptors, waterfowl and other large birds, electrocuted when their wings bridge two hot wires.

Another 100 million to 900 million birds perish after flying into windows, mostly in rural areas, according to another report.

Pesticides kill an estimated 67 million birds each year. Scientists estimate that about 60 million birds die each year in the United States after being struck by motorized vehicles, according to the American Wind Energy Association's report "Facts about Wind Energy and Birds."

Yet another 40 to 50 million birds perish after flying into communications towers and the guy wires that support them. Studies of one television transmitter tower in Eau Claire, Wisconsin, showed that it killed over 1,000 birds a night on 24 consecutive nights. This same tower killed a record 30,000 birds one evening! A similar tower in Kansas killed 10,000 birds in a single evening.

Another 1.25 million die as a result of collisions with tall structures such as buildings, smokestacks and towers.

Clearly, the Altamont Pass wind farm is benign compared to a host of other lethal factors. Altamont is also an isolated case. No other wind farm in the United States experiences mortality rates remotely close to Altamont Pass. Why?

Contemporary wind developers have been selecting sites for new wind farms that are out of migratory pathways. Improvements in the design of commercial wind turbines have also helped to minimize bird kills at commercial wind farms. Over the years, wind machines have gotten taller, blades have gotten longer, and the speed at which the blades rotate has declined substantially. These large, slow-moving blades are more easily avoided by birds. Ever-larger commercial wind machines currently under development could reduce the risk even more.

As shown by the graph below, large commercial wind turbines are an extremely minor source of bird mortality.

"Double the number of turbines," Mick adds, "and we're up to 0.02 to 0.04 percent. Increase the number by 1,000 percent and we're up to 0.1 percent! How many birds does habitat destruction such as mountain top removal to mine for coal kill — forever?!?"

To learn more about efforts to further reduce bird deaths, check out "Facts about Wind Energy and Birds" at the American Wind Energy Association's website, awea.org. Mick has written several articles on the topic, which you can check out at renewwisconsin.org. On the lower left, click on "Small Wind Toolboxes."

Aesthetics

Another downside of wind turbines is that some people don't like the look of them or believe that wind turbines detract from natural beauty. While some individuals object to the sight of a residential wind turbine or a commercial wind farm, others find them to be things of great beauty. Ironically, those who find wind turbines to be unsightly often ignore the great many forms of visual blight that litter our landscape, among them cellphone towers, water towers, electric transmission lines, radio towers and billboards. To be fair, there are differences between a wind tower and common sources of visual pollution. For one, wind turbines with their spinning blades call attention

to themselves. Another is that we've grown used to the ubiquitous electric lines and radio towers. As a result, many often fail to see them anymore. Yet another difference is that the structures we ignore are often ones that were erected without public consent. That is, they were exempt from public hearings. People had no choice in their placement and have eventually grown used to them.

Given the opportunity to oppose a structure in their viewshed — for example, at a public hearing required for permission to install a residential wind system — Mick points out that many neighbors are quite willing to speak up in opposition. As he notes in an article on aesthetics in AWEA's *Windletter*, "anyone who has tried to deal with aesthetics in a public

Fig. 1-7: *This graph shows the relative number of bird deaths from various sources. Note that wind turbines are responsible for only a tiny portion of total annual bird deaths.*

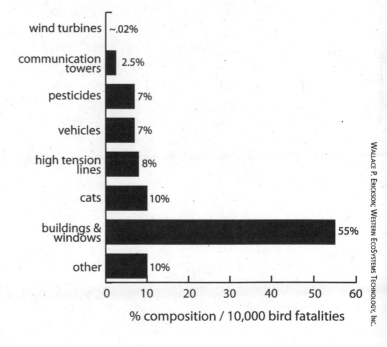

WALLACE P. ERICKSON; WESTERN ECOSYSTEMS TECHNOLOGY, INC.

hearing knows only too well why art has never been created by committee."

While the battle continues over commercial wind development and individual battles arise as homeowners or business owners attempt to install small-scale wind to meet their needs, it may be comforting to those who support wind to learn that when windmills were first introduced into Holland, they were looked upon by some with distaste. Another case in point: a large commercial-sized wind machine recently built at the Portsmouth Abbey School in Rhode Island drew criticism at first, but is now widely loved by residents of the community. In Chapter 10, we'll discuss ways individuals can help prevent and overcome opposition from neighbors based on aesthetics.

Proximity to Homes and Property Values

Critics do raise legitimate concerns when it comes to the placement of wind machines near their property. Although most of the issues over proximity have been raised by individuals and groups that oppose large commercial wind farms, residential systems can also cause a stir among neighbors. Some may be concerned about aesthetics. Others may worry about safety.

To avoid problems, we recommend installing machines, whenever possible, in locations out of sight and hearing of sensitive neighbors. Although tower collapse is an extremely rare event and always the result of bad design and improper installation and homeowner's insurance should cover damage to individuals and property, it is best to place a wind turbine and tower away from your neighbors' property lines. This could help overcome objections.

Unwanted Sound

Opponents of wind energy and apprehensive neighbors sometimes voice concerns about unwanted sound, aka noise, from residential wind machines. Small wind turbines do produce sound and as the wind speed increases, sound output increases.

Sound is produced primarily by the spinning blades and alternators. The faster a turbine spins, the more sound it produces.

An Opposing View

"I think that the goal should be to site wind turbines near homes without fear," Ian argues. "It's better to build relationships with your neighbors, and get them excited about renewable energy. Then the question will be 'Will I be able to see the blades spin?' not 'Must I look at it?' Some wind energy users," he adds, "share electricity with neighbors during utility outages, which builds interest and appreciation." Working out an arrangement like this in advance may help overcome barriers with reluctant or skeptical neighbors and win their support. When dealing with neighbors, remember that many of them share your excitement for renewable energy. Many people would love to reduce their electric bill or achieve greater energy independence, or just enjoy the coolness factor that owning a wind turbine brings, but not all can. "Your neighbors can be cool by association," Ian notes, "if you do the PR in advance."

Individuals can reduce unwanted sound by selecting quieter low-rpm wind turbines. As we point out in Chapter 5, high-rpm wind turbines tend to be louder than low-rpm units. If you are concerned about sound, make this a high priority as you shop for a turbine and let your neighbors know this is an issue to which you are sensitive.

Wind turbines also come with governing mechanisms, systems that slow down or even turn off the machines, when winds get too strong, to protect them from damage. Different governing systems result in different sound levels. (We'll discuss this topic in Chapter 5.) When researching your options, we recommend that you listen to the turbines you're considering buying in a variety of wind conditions, including those that require governing.

Besides buying a quieter wind turbine to reduce sound, it's also important to mount your turbine on a tall tower. Suitable tower heights, which we'll discuss later, are usually at least 80 to 120 feet. A residential wind turbine mounted high on a tower catches the smoother and stronger — and hence most productive — winds. This strategy also helps reduce sound levels on the ground. Part of the reason for this is that sound dissipates quickly over distance. (For mathematically inclined readers, sound decreases by the square of distance.)

Residential (and commercial) wind machines are also much quieter than many people suspect because the sounds they make are partially drowned out by ambient sounds on windy days. Rustling leaves and wind blowing around one's ears often drown out some of the sound produced by a residential wind turbine.

Sound is measured in two ways — by loudness and frequency. Loudness is measured in decibels (dB). Frequency is the pitch. A low note sounded on a guitar has a low frequency or pitch. A high note has a high frequency. Interestingly, the average background noise in a house is about 50 dB. Nearby trees on a breezy day measure about 55 to 60 dB. According to Mick, "Most of today's residential wind turbines perform very near ambient levels over most of their effective operating range." However, even though the intensity of a sound produced by a wind generator may be the same as ambient sound, the frequency may differ. As a result, wind turbine sounds may be distinguishable from ambient noises, even though they are not louder.

"Today's home-sized wind turbines typically operate from just below to just above ambient environmental sound levels at their loudest when governing," he adds. "This means that while the sound of a wind turbine can be picked out of surrounding noise if a conscious effort is made to hear it, home-sized wind turbines are by no means the noisy contraptions that some people make them out to be."

For more on sound, you may want to read Mick's article, "Residential Wind Turbines and Noise" in the April 2004 issue of AWEA's *Windletter*. We'll also spend more time on this topic in Chapter 5 on wind turbines and will discuss strategies for addressing sound issues at zoning hearings in Chapter 10.

"Remember that sound is very subjective — what to some might be irritating is the pleasant sound of renewable energy at work to others."

—Ian Woofenden

Fig. 1.8: *Perched on top of this 168-foot tower is wind energy expert, author, and workshop teacher Ian Woofenden who served as a primary technical advisor on this book. This extremely tall tower raises the turbine well above the trees that carpet the island where Ian lives, allowing access to the wind and permitting excellent performance.*

Site Specific

Yet another criticism of wind that's of great importance to small wind systems is that it is more site specific — or restricted — than solar energy.

To understand what this means, we begin by pointing out that there are good solar areas and good wind areas. In a good solar region, most people with a good southern exposure can access the same amount of sun. In a windy area, however, hills and valleys or stands of trees can dramatically reduce the amount of wind that blows across a piece of property. Therefore, even if you live in an area with sufficient winds, you may be unable to tap into the wind's generous supply of energy because of topography or vegetation like tall stands of trees. That's what critics mean when they say that wind energy is more site specific.

That said, we'd be remiss if we did not point out that solar resources also vary. If you live in a forest, you'll have less solar energy than a nearby neighbor whose home is in an open field. In addition, we should point out that homeowners can access the wind at less-than-optimum sites by installing turbines on tall towers. Ian, for example, recently installed an ARE110 turbine, made by Abundant Renewable Energy in Oregon, in a densely forested island in the Pacific Northwest. He made it work by installing the turbine on a 168-foot tower well above the tops of the trees. Tall towers help us overcome topographical and other barriers.

We'd also be remiss not to mention that wind can be augmented or "magnified" at a site. That is, an individual can harvest more wind energy at a site by increasing tower height. As Mick points out in his wind energy workshops, you can't make a location sunnier, but by increasing tower height you can move a turbine into higher velocity winds and achieve much greater output.

Ice Throw

Like trees and powerlines, wind turbines can ice up under certain conditions. Ice buildup and possible ice throw and the dangers they pose are issues that may arise during hearings on residential wind turbines.

While ice builds up on blades in ice storms, it is typically deposited on turbines and towers in very thin sheets. When the blades are warmed by sunlight, however, the ice tends to break up into small pieces, not huge and potentially dangerous chunks.

Ice buildup on the blades of a wind turbine also dramatically reduces the speed at which a turbine can spin.[2] It's a little like trying to drive a car with four flat tires. As a result, ice is not thrown great distances; it tends to fall around the base of the tower — just as it does from trees and power lines.

Any prudent person would be advised to stay away from the tower base when ice is shed from the blades, as they would from ice falling from trees or power lines. Ice-laden trees are also considerably more dangerous, as ice-coated branches can and often do break and fall to the ground, damaging power lines and cars or houses. Entire trees can topple as a result of ice buildup.

On the rare occasion that ice builds up on a wind turbine, experienced wind turbine operators shut down their machines until the sun or warmer temperatures melt the ice since they cannot generate electricity spinning at such low rpms anyway.

Interference with Telecommunications

Some opponents of wind energy also raise the issue of interference with telecommunications signals.

While there are a few reports of large-scale wind turbines causing interference with television reception, these problems arose because the turbines were installed directly in the line of sight between the TV transmitter and a residential antenna. The spinning blades chopped up the signal, causing flickering on televisions. Interference represented isolated cases and was easily corrected by installing larger antennas or signal boosters.

With small wind turbines, interference is extremely unlikely. Turbines for homes and small businesses have small blades that do not interfere with such signals. The blades of modern wind turbines are also made out of materials that are unlikely to cause problems. Unlike the metal blades of years past which can reflect TV signals, the fiberglass and plastic blades in use today are "transparent" to telecommunications signals.

As a case in point, we should note that small wind turbines are often installed to power remote telecommunications sites. The US-based company, Abundant Renewable Energy, which manufactures two wind turbines, mounted their Internet receiver/transmitter on their wind turbine tower. Telecommunication equipment wouldn't be installed in such locations if there was a problem with interference.

Photoepilepsy and Strobing

Yet another issue that may be raised from time to time by concerned neighbors or opponents of wind energy is the possibility of shadow flicker from wind turbines stimulating epileptic seizures in individuals who suffer from photosensitive epilepsy. This is an extremely rare type of epilepsy in which seizures are triggered by flickering or flashing light.

This concern, most often raised by opponents of large commercial wind turbines, but occasionally raised at zoning hearings for small wind turbines, is not just overblown, it's not even true. According to researchers, there's never been a case of epileptic seizure triggered by a wind turbine in human history.

While blades of small turbines may form small and rather vague shadows, it is difficult to see the shadow of individual blades due to the speed with which the blades spin. As Jim Green notes, "The rotors of residential-scale wind turbines, 10 kilowatts and smaller, essentially become transparent at typical operating speeds because the blades spin faster than the eye can detect."

The true test of this issue's serious may come from installers and dealers. Mike Bergey of Bergey Windpower, for example, has never received a complaint about shadow flicker from customers or neighbors of the more than 3,000 turbines his company has sold over more than two decades.

Property Values

Opponents of wind farms often raise the specter of declining property values, despite the lack of any evidence to support their assertions. Nonetheless, concerns over property values often arise in zoning hearings over small wind turbines. As Mick puts it, the rationale is that the neighborhood viewshed will be compromised as a result of the installation of a home-sized wind turbine. Neighbors worry that they will not be able to sell their property for its true value.

While wind turbines on tall towers are visible, lots of other tall structures like silos, barns, high-power transmission lines, water towers and cell phone towers are present in rural environments where residential and small business wind systems are typically installed. Small wind systems are often much less visible than these structures.

Moreover, we've never heard of an instance in which a residential wind turbine adversely affected the value of a neighbor's property. For the system owner, a wind turbine could increase property values, in part as a result of reduced utility bills.

The Advantages of Wind Energy

Although residential wind turbines and their energy source, the wind, have their downsides, many features make them well worth considering. To begin with, wind energy is an abundant and renewable resource. We won't run out of wind for the foreseeable future, which stands in stark contrast to the future of oil and natural gas.

Small-scale and large-scale wind energy could help decrease our reliance on declining and costly supplies of oil. Electricity generated by wind, for instance, could be used to power electric or plug-in hybrid cars and trucks in areas with abundant wind resources, displacing gasoline, which is refined from oil.

Wind energy — both large and small — can also play a meaningful role in offsetting declining US natural gas supplies. In the United States, approximately 18 percent of all electricity is currently generated by natural gas, according to the US Department of Energy. As supplies continue to decline, wind could help ease the crunch, supplying a growing percentage of our nation's electrical demand long into the future.

Wind could even replace nuclear power plants the world over. Nuclear power plants generate about 20 percent of America's electricity, and substantially higher percentages in countries such as France. Although wind energy does have its impacts, it is a relatively benign

100+ MPG

CALCARS.ORG

CALCARS.ORG

technology compared to fossil fuel and nuclear power plants. Because of this, it could help all countries create a cleaner and safer energy future at a fraction of the cost and impact of conventional electrical energy production.

Another benefit of wind energy is that, unlike oil, coal and nuclear energy, the wind is not owned by major energy companies. The cost of wind is not subject to price increases, protecting us from price gouging by commodity traders and multinational corporations. Price hikes caused by rising fuel costs are not probable in a wind-powered future. However, this is not to say that wind energy is immune to the rising price of fossil fuels. As Ian notes,

while the fuel itself (the wind) will not increase in price, the price of wind generators is likely to increase as traditional fuel prices rise. That's because it takes energy to extract and process minerals to make the steel and copper needed for wind turbines and towers. It also takes energy to make turbines and towers and ship and install them.

Another huge benefit to consider is that wind could also be used to power electric and plug-in hybrid vehicles, helping nations wean themselves from their costly oil addiction, clean up their air, and halt global warming.

An increasing reliance on wind energy could also ease political tensions worldwide. If

Fig. 1.9: *Electric cars and plug-in hybrids like the one shown here are the most promising automobile technologies on the horizon.*

we free ourselves from Middle Eastern oil we won't need costly military operations aimed, in part, at stabilizing a region where the largest oil reserves reside. We'll likely never fight a war over wind energy resources. Not a drop of human blood need be shed to ensure a steady supply of wind energy to fuel the economy.

Yet another advantage of wind-generated electricity is that it uses existing infrastructure, the electrical grid, and existing technologies like electric toasters, microwaves and so on. A transition to wind energy could occur fairly seamlessly.

Individuals can also meet all or part of their energy needs in rural areas with good wind resources at rates that are competitive with conventional electricity. In remote locations, wind or wind and solar electric hybrid systems may be cheaper than conventional power delivered through newly installed and costly electric lines from the utility grid.

Finally, lest we forget, wind is a clean resource. Wind energy will help homeowners and businesses do their part in solving costly environmental issues such as acid rain and global climate change. As Mick points out in his workshops, the average home in the United States consumes 900 kilowatt-hours of electricity per month. Replacing the electricity generated by a coal-fired power plant with wind-generated electricity will reduce a family's consumption of coal by approximately 5.5 tons per year. This,

Fig. 1-10: *Wind turbines like these in central Kansas are typically sited to reduce mortality and, as explained in the text, are responsible for only a tiny portion of total annual bird kills. The real culprits are cats, buildings, cell phone towers, cars and trucks, and pesticides.*

DAN CHIRAS

in turn, will reduce the emission of carbon dioxide by about 11 tons per year. It will also reduce mercury emissions. You couldn't ask for more reasons to justify a switch to wind energy.

Wind energy also provides some substantial economic benefits. In 2007, for instance, nine billion dollars was invested in US wind farms, according to the American Wind Energy Association. Wind also creates more jobs per kilowatt-hour generated than other type of power plant. It also concentrates economic benefits locally, within states or communities. And wind power does not require extensive use of water, an increasing problem for coal, nuclear and gas-fired power plants, particularly in the western US and other drought-stricken areas.

The Purpose of this Book

This book's principal focus is on small wind-electric systems. As noted earlier, the rated output of small wind turbines ranges from 1 kilowatt to 100 kilowatts. Most of the turbines we'll be discussing fall in a range from 1 kilowatt to 20 kilowatts. The blades of small turbines (1 to 100 kilowatts) run from 4 feet to 32 feet in length. Small-scale wind systems serve a variety of purposes. The smaller units are sufficient to power cabins and cottages and larger turbines power homes, small businesses, schools, farms, ranches, manufacturing plants and public facilities. Throughout this book, we'll refer to these applications as small wind systems or small-scale wind systems, sometimes even residential and business wind energy systems, to avoid having to repeat the long list of applications.

This book is written for individuals who want to learn about small-scale wind systems. It is also written for those who aren't particularly well versed in electricity and electronics. You won't need a degree in electrical engineering, renewable energy, or physics to make sense of the material covered in this book.

The overarching goal in writing this book was to create a user-friendly book that teaches readers the basics of wind energy and wind energy systems. We should point out emphatically that this book is *not* an installation manual. It will not turn you into a wind energy installer or equip you to install a wind turbine and tower on your own. It will, however, help you determine if wind energy is right for you. When you finish reading and studying the material in

"Envision a future in which distributed (small) wind power is *embraced* in the local landscape because it *expresses community support* for clean air, reduced carbon emissions and strong local economies through use of a sustainable, indigenous energy source."

— Jim Green, National Renewable Energy Laboratory

Fig: 1-11: *Small wind turbines like this one from Canada's True North Power Systems are rarely, if ever, responsible for bird deaths.*

TRUE NORTH POWER SYSTEMS.

this book, you'll know an amazing amount about wind and wind energy systems. You will have the knowledge required to assess your electrical consumption as well as the wind resource at your site, to determine if wind will meet your needs.

When you are done with this book, you should have a good working knowledge of the key components of wind energy systems, especially wind turbines, towers, batteries and inverters. In keeping with our long-standing goal of creating knowledgeable buyers, this book will help you know what to look for when shopping for a wind energy system. You'll also know how wind machines are installed and their maintenance requirements.

If you choose to hire a professional wind energy expert to install a system, a route we highly recommend, you'll be thankful you've read and studied the material in this book. The more you know, the more input you will have into your system design, components, siting and installation — and the more likely that you'll be happy with your purchase.

In keeping with another long-standing goal of ours, this book should also help readers develop realistic expectations. We believe that those interested in installing renewable energy systems need to proceed with their eyes wide open. Knowing the shortcomings of wind energy — or any renewable energy technology, for that matter — helps avoid mistakes and prevents disappointment often fueled by unrealistic expectations. Wind energy systems, for instance, require annual inspection and maintenance — climbing or lowering a tower to

access the wind turbine to check for loose fasteners and blade damage and, much less commonly, an occasional part replacement. If you are not up for it or don't want to pay someone to climb or lower your tower once or twice a year to check things out, you may want to invest in a solar-electric system instead.

Organization of this Book

Now that you know a little bit about the history of wind energy, the pros and cons of this clean, renewable energy source, and the purpose of this book, let's start our exploration. We'll begin in the next chapter by studying wind, the driving force in a wind energy system. You will learn how winds are generated and explore the factors that influence wind flows in your area.

In Chapter 2, we will also explore the factors that affect energy production by a residential wind turbine. We call this the mathematics of wind energy. The math isn't difficult, and this discussion will demonstrate how the proper design and placement of a wind machine can result in dramatic increases in electrical output. When you finish, you will understand why it is important to mount a wind machine as high as you can and out of the way of obstructions that reduce wind speed and create turbulence when the winds blow. This advice could make the difference between a successful wind venture and a costly failure.

In Chapter 3, we'll explore wind energy systems. You'll learn the three types of residential wind energy systems: (1) off-grid, (2) batteryless grid-tie, and (3) grid-connected with battery backup. You'll also learn about the basic components of each one. We'll also look at hybrid wind systems.

Chapter 4 explores the feasibility of tapping into wind at your site. We'll teach you how to assess your electrical energy needs and how to determine if your site has enough wind to meet them. You'll learn why cost-effective energy efficiency measures that reduce your electrical demands will save you heaps of money when buying a wind energy system. You'll also learn ways to evaluate the economics of a wind system.

Chapter 5 introduces you to wind turbines. You'll learn about the different types of wind turbines and how they work. We'll also give you shopping tips — what to look for when buying a wind turbine. We'll even spend a little time looking at ways to build your own wind generator.

Chapter 6 describes three basic tower options. You will learn how towers and guy wires (used to support certain types of towers) are anchored. We'll underscore the importance of mounting a wind machine high above the ground — out of turbulent ground-level air and dead air zones and into the much smoother and more powerful winds that blow higher up. We'll also look briefly at how towers are installed.

In Chapter 7, we'll tackle storage batteries, one of the key components of off-grid wind systems. You will learn whether you will need a battery bank and, if so, what kind of batteries you should install. You will learn about battery care and maintenance and ways to make your life with batteries much easier. You'll benefit from our combined decades of experience with battery systems as we point out common mistakes and ways to avoid them. You will also learn about battery safety and how to size a battery bank for a wind energy system.

Chapter 8 addresses another key component of wind energy systems, the inverter. You will learn how inverters work, what functions they perform, and what to look for when shopping for one.

In Chapter 9, we'll give a brief overview of wind energy system maintenance. Because each wind machine and tower is different, we won't go into specifics. We will, however, underscore the importance of regular inspection and maintenance and describe some of the most common things you'll have to do to keep your wind energy system performing optimally.

With this information in mind, in Chapter 10 we'll explore a range of issues such as homeowner's insurance, financing renewable energy systems, obtaining building permits and electrical permits, and zoning issues.

Finally, this books ends with a fairly comprehensive resource guide. It contains a list of books, articles, videos, associations, organizations, workshops and websites on residential wind energy.

What do you say — shall we get started?

CHAPTER 2

UNDERSTANDING WIND
AND WIND ENERGY

Wind can be an asset and a liability. Growing up along the windswept shores of Lake Ontario in western New York, fierce winter winds rattled the aluminum storm windows of my bedroom, waking me in the middle of the night. Other times, the winds created nightmarish blizzards that made driving impossible.

Wind has proved to be a liability in my adult years, too. On several kayak trips on western rivers like the Colorado and Green Rivers, the winds were so strong that they flipped us over while paddling in relatively calm water. When kayaking the Grand Canyon, powerful winds sometimes halted the forward motion of the support rafts that carried our gear. Several of us had to climb aboard the rafts to help the oarsmen battle the winds in calm sections. And on several of my backpacking trips in the high peaks of Colorado, powerful winds

have made travel across the tundra nearly impossible. Sometimes, it seemed as if we'd be blown off the mountaintops.

Wind has also been a blessing. Those powerful winter winds off Lake Ontario, for instance, created huge snowdrifts along the hedgerow that bordered my family's property. Undaunted, my brothers and I would dig through the deep drifts with snow shovels, creating a network of interconnecting tunnels where we could hang out in safety, protected from the fierce winds.

Today, wind continues to bless my life. It provides a portion of the electricity to my solar home and office in the foothills of the Rocky Mountains. Wind will also provide electricity to power my retirement home and renewable energy and green building education center in eastern Missouri. Wind is also an asset to my colleagues, Mick Sagrillo and Ian Woofenden, the two wind experts who've joined me in writing this book. In Mick's home in Wisconsin, wind turbines provide 100 percent of his and his wife's electrical demands. Over the years, Ian has generated as much as 60 percent of his family's electricity for his island home in the Pacific Northwest from the wind.

This chapter will introduce you to the wind — the driving force of a wind energy system. We'll examine how wind is generated and study four distinct phenomena: (1) offshore and onshore winds, (2) mountain-valley winds, (3) global circulation patterns, and (4) storms. We'll also explore ways local topography affects wind, introducing you to two key concepts in wind energy: ground drag (friction) and turbulence. Chapter 4 will cover more details on wind, including the influence of topography and vegetation on wind.

As you read this material, bear in mind that this is not just a theoretical exercise. It provides the practical knowledge you will need to select the best site for a wind turbine and the optimum tower height.

What is Wind?

Wind is air in horizontal motion across the Earth's surface. All winds are produced by differences in air pressure between two adjoining regions. Differences in pressure result from differential heating of the surface of the Earth, as you shall soon see. Heating, of course, is the work of the sun.

Like most other forms of energy in use today, even coal, oil and natural gas, wind is a product of sunlight — pure solar energy. Some wind advocates, in fact, refer to wind as "the other solar energy," along with hydropower (the hydrologic cycle is also driven by the sun). Robert Preus of Abundant Renewable Energy, which manufactures residential wind turbines, calls wind "second hand solar energy." Paul Gipe, an author of several books on wind energy, moderated a workshop at Solar 2007, the annual conference of the American Solar Energy Society appropriately titled "Bringing 'Solar Convection' Technology into the Mainstream: Small Wind Turbines."

To understand wind, we begin by looking at two types of local winds: (1) offshore and onshore winds and (2) mountain-valley breezes.

Offshore and Onshore Winds

Offshore and onshore winds are, as their names imply, generated along coastlines. That is, they occur along the shores of large lakes, such as the Great Lakes of North America, and along the coastlines of the world's oceans. Offshore and onshore winds blow regularly, nearly every day of the year. Like all other types of wind, solar energy generates these winds. How does sunlight create offshore and onshore winds?

As shown in Figure 2.1a, sunlight shining on the Earth's surface heats the land and nearby water bodies. As the water and adjoining land begin to warm, they radiate some of the heat (infrared radiation) into the atmosphere. This heat, in turn, warms the air immediately above them. When air is heated it expands, and as it expands it becomes less dense. As the density of the air decreases, it rises. The upward movement of air is called a thermal or updraft.

Although water and land both heat up when warmed by the sun, land masses warm more rapidly than neighboring bodies of water. Because air over land heats up more quickly than air over water, air pressure over a land-mass is lower than it is over neighboring surface waters.

Low-pressure air rising from the land creates something of a void, which is filled by cooler, higher-pressure air lying over neighboring water. As the cool air flows in to fill the void, a breeze is created. The breeze is known as an onshore breeze or onshore wind.

At night, the winds blow in the opposite direction — that is, they blow from land to water, as illustrated in Figure 2.1b. These are known as offshore breezes or offshore winds.

Like onshore winds that occur during the day, offshore winds are created by differences in air pressure between the air over land and neighboring water bodies. Differential cooling generates these differences. Here's what happens: after sunset, the land and the ocean both begin to cool. However, land gives up heat more rapidly than water. As a result, the land surfaces cool more rapidly than nearby water bodies. As shown in Figure 2.1b, at night the air over water continues to release heat, causing the air above it to expand and rise. As the warm, low-pressure air rises, it creates a slight updraft over the water. Cooler, higher-pressure air from land flows in to fill the vacuum. The result is an offshore breeze: steady winds that flow from land to water.

In the absence of prevailing winds or storms, discussed shortly, offshore and onshore breezes operate day in and day out on sunny days, providing a steady supply of wind energy to power wind turbines. Because offshore and

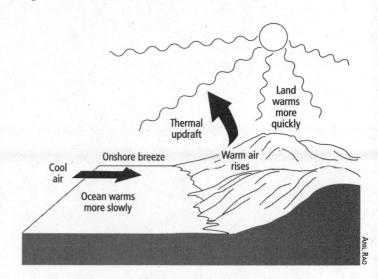

a

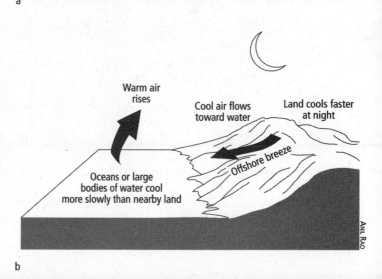

b

Fig. 2.2: *Onshore and offshore breezes. Onshore (a) and offshore (b) breezes occur along the coastlines of major lakes and oceans.*

Fig. 2.2: *Mountain-Valley Breezes. Mountain-valley winds can provide a reliable source of wind power if conditions are just right. (a) Up-valley winds. (b) Down-valley winds.*

onshore winds are fairly reliable, coastal regions of the world are often ideal locations for small wind turbines.

Coastal winds are a great energy resource and could help many countries meet their needs for electricity. In fact, it is not unusual to encounter winds with an average annual speed of 12 to 18 miles per hour (5 to 8 meters per second) in such locations. Wind speeds such as these are ideal for residential and commercial wind generators, although the onshore

and offshore wind effect is restricted to within a mile or so from the coastline.

Coastal winds are not only more consistent than winds over the interior of continents, they also tend to be more powerful because of the relatively smooth and unobstructed nature of the surface of open waters. Put another way, wind moves rapidly over water because lakes and coastal waters provide very little resistance to its flow, unlike forests or cities and suburbs, which dramatically lower surface wind speeds.

Mountain-Valley Breezes

Like coastal winds, mountain-valley breezes arise from the differential heating of the Earth's surface. To understand how these powerful winds are formed, let's begin in the morning.

As the sun rises in the eastern sky on clear days, sunrays strike the valley floor and begin heating the ground, valley walls and mountains. As the ground and valley walls begin to warm, the air above them warms. It then expands and begins to flow upward. This process is known as convection. (Convection is the transfer of heat in a fluid that is caused by the movement of the heated fluid itself. In a building space, warm air rises and cold air settles to create a convection loop.) While some of this air rises vertically, mountain valleys also tend to channel the solar-heated air up through the valley toward the mountains (Figure 2.2a). As the warmed air moves up a valley, cooler air from surrounding areas such as nearby plains flows in to replace it. This wind is known as a valley breeze.

Throughout the morning and well into the afternoon, breezes flow up-valley — from the

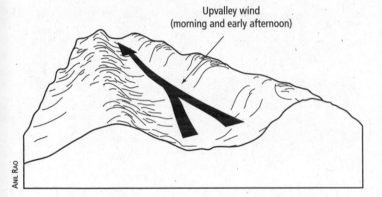

Upvalley wind
(morning and early afternoon)

Anil Rao

a

Downvalley wind
(late afternoon and evening)

Anil Rao

b

valley floor into the mountains. These breezes tend to reach a crescendo in the afternoon. When the sun sets, however, the winds reverse direction, flowing down valley.

Winds flow in reverse at night because the mountains cool more quickly than the valley floor. Cool, dense air (high-pressure air) from the mountains sinks and flows down through the valleys like the water in mountain streams, creating steady and often predictable down-valley or mountain breezes.

Mountain breezes typically start up shortly after sunset and continue throughout the night, terminating at sunrise. As a rule, mountain winds are generally strongest in valleys that slope steeply and valleys cradled by high ridges. The strongest mountain winds occur at the mouth of the valley (that is, the lower reaches). The winds in the upper reaches of the valley also tend to be smoother and less turbulent than the winds in the lower parts.

Together, valley and mountain winds are known as mountain-valley breezes. As a rule, mountain breezes (down-flowing winds) tend to be stronger than daytime valley breezes. Mountain breezes often reach speeds around 25 miles per hour (about 11 meters per second), ideal for small and large-scale wind energy production. Wind flow is most rapid, as a rule, in the center of the valley — at about two-thirds the height of the surrounding ridges.

Mountain-valley breezes typically occur in the summer months, a time when solar radiation is greatest. They also typically occur on calm days when the prevailing winds (larger regional winds, discussed shortly) are weak or nonexistent.

Mountain-valley winds may also form in the presence of prevailing winds — for example, when a storm moves through an area. In such instances, mountain or valley winds may "piggy back" on the prevailing winds, creating even more powerful (and hence higher energy) winds. When consistently flowing in the same direction, such winds can provide a great deal of power that can be tapped to produce an abundance of electricity.

When assessing a wind site in a mountain valley, the orientation of the valley to the prevailing wind direction is extremely important. If the prevailing winds run in the same direction as the valley floor, they can enhance the wind resource — and could dramatically increase the amount of electricity that can be generated by a wind turbine.

When looking for land to build on and install a wind turbine, shop carefully in mountainous terrain. The ideal property in such terrain is a valley that runs parallel to the prevailing wind in the summer, when mountain-valley winds are strongest. In such locations, mountain-valley winds can piggyback on the prevailing winds, creating more harvestable wind power. In the winter, when the mountain-valley breezes are the weakest, prevailing winds will continue to flow through the valley, spinning the blades of a wind turbine. (Bear in mind that winter and summer winds may come from different directions.)

As a rule, "a valley that is both parallel to the prevailing wind and experiences mountain-valley winds will provide sites that are dependable sources of power," according to the authors of *A Siting Handbook for Small Wind Energy Conversion Systems*. If the valley narrows at a site, the narrowing creates a funnel effect (aka the Venturi effect) that further increases wind speed — and the potential output of a wind generator.

If the prevailing winds run perpendicular to the valley floor, or at an angle greater than 35 degrees, expect much less electrical energy from a wind turbine. Such winds tend to flow over valleys. The deeper and narrower the valley, the more likely cross-valley winds will flow over it.

Cross-valley winds may also combine with solar heated air rising from the valley floor to create eddies — swirling turbulent air. As the authors of *A Siting Handbook for Small Wind Energy Conversion Systems* point out, this is not a reliable source of energy. Furthermore, as you shall soon see, turbulence wreaks havoc on wind turbines.

Global Air Circulation

The flow of air between the equator and north and south poles distributes excess heat from equatorial regions throughout the planet. (As a result the equator is cooler and the temperate and polar regions are warmer than they would be without global air circulation.) This massive movement of air is also the source of winds that could supply human society with a significant portion of its electrical energy needs.

When considering a wind site in a valley, bear in mind that not all valleys are created equal. Valleys that do not slope downward from mountains are not usually good wind sites, unless they're in regions with extremely strong winds or they narrow significantly, funneling prevailing winds through them. (The Columbia Gorge is a good example of the latter.) If you're thinking about installing a wind turbine in a valley not connected to a chain of mountains, you'd probably do better placing it on a surrounding ridge.

Large-Scale Wind Currents

Local winds can be a valuable source of energy for homes and businesses powered by wind generators. However, the winds on which most people rely are those derived from much larger air flows resulting from regional and global air circulation. They create dominant wind flow patterns, known as prevailing winds.

Prevailing winds, like local winds, are created by the differential heating of the Earth's surface, but on a much larger scale. Here's a simplified version of how they are created. As shown in Figure 2.3, the Earth is divided into three climatic zones: the tropics, temperate zones and poles. As illustrated, the tropics straddle the equator, extending from 30° north latitude to 30° south latitude. Because the tropics are more directly aligned with the sun throughout the year than are other zones, they receive considerably more sunlight. As a result, the tropical region is the warmest region on Earth.

The temperate zones lie outside the tropics, both in the Northern and Southern

Hemispheres. They are located between 30° and 60° north latitude in the Northern Hemisphere and 30° and 60° south latitude in the Southern Hemisphere. They receive less sunlight than the tropics and are, therefore, noticeably cooler. Last but not least are the North and South Poles. They receive the least amount of sunlight and are the coolest regions of our planet, and historically have been covered by ice much of the year. (That's changing because of global warming.)

The unequal heating of the Earth's surface is responsible for the movement of huge air masses across its surface. As shown in Figure 2.4a, a highly simplified view of solar convection, hot air produced in the tropics rises upward. Cool air from the northern regions — as far north as the poles — moves in to fill the void. The result is huge air currents that flow from the poles to the equator.

Although air generally flows from the North and South Poles toward the equator, circulation patterns are much more complicated. In the Northern Hemisphere, for instance, warm air from the equator rises and moves northward. As shown in Figure 2.4b, as the warm tropical air, which is laden with moisture, flows northward it begins to cool, depositing moisture. (Most of the moisture falls back on the Tropics, being locally recycled.)

As shown in Figure 2.4b, some of the cooler northbound air sinks back to the Earth's surface and flows back toward the equator, creating a short-circuit in global wind circulation. This results in winds that blow toward the equator, known as the trade winds.

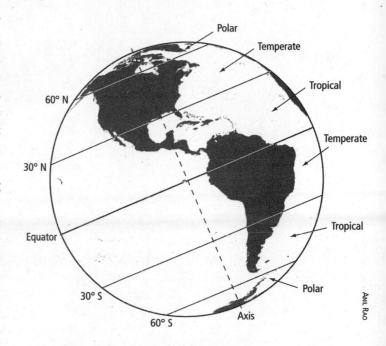

Because the trade winds blow quite consistently, day after day, they are a potentially huge and reliable source of energy, a fact not missed by sailors who have relied on them for centuries. (The trade winds got their name from their importance to trade in the early days of commerce.) For residents and nations fortunate enough to lie in the trade winds, these constant winds could become a valuable source of electrical energy.

As shown in Figure 2.4b, not all of the air flowing northward from the equator returns via the trade winds. A substantial amount of the northward moving air proceeds toward the North Pole. Devoid of much of its moisture, the relatively dry air travels over the temperate zone.

Fig. 2.3:
Climate Zones.
The Earth is divided into three climate zones in each hemisphere. In the Northern Hemisphere, warm air from the tropics flows northward by convection, creating the global circulation pattern shown here.

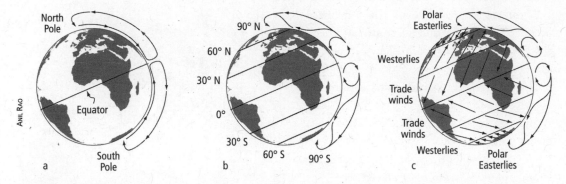

Fig. 2.4: *Global Air Circulation. (a) As shown here, warm tropical air rises and flows toward the poles. Cold polar air flows toward the equator. (b) Air circulation is more complicated than shown in (a), however. Warm tropical air loses some of its heat and sinks, creating a circulation pattern that brings air back toward the equator, creating the trade winds. Air masses moving over the temperate zone split into upper and lower winds. (c) Wind patterns caused by the Coriolis effect, resulting from the rotation of the Earth on its axis.*

As Figure 2.4b also shows, the air masses flowing northward across the temperate zone splits into higher- and lower-level winds. As it flows northward, the lower-level air picks up a considerable amount of moisture, much of which falls as rain and snow over the temperate zone.

When the equatorial air reaches the North Pole, however, it is dry and cold. This cold, dry air mass then sinks and begins to flow southward back toward the equator.

If no other forces were at work, winds flowing back to the equator would flow from north to south. As shown in Figure 2.4c, they don't. Other factors influence the movement of air masses across the surface of the planet. One of the most significant is the Earth's rotation, which results in a phenomenon known as the Coriolis effect, described next.

The Coriolis Effect

In order to understand why prevailing winds deviate from the expected flow patterns based solely on convection, let's look at the Northern Hemisphere once again. We'll focus our attention first on the trade winds. As shown in Figure 2.4c, the trade winds in the Northern Hemisphere flow not from north to south, as you might expect, but from the northeast to southwest. Why?

Because they are "deflected" by the Earth's rotation.

In reality, the Earth's rotation doesn't deflect winds. It makes it appear as if the winds have been deflected. The apparent deflection in wind direction in the tropics is a planetary sleight of hand, an illusion produced by the rotation of the Earth on its axis. To understand this phenomenon, let's look at a simple example. Imagine

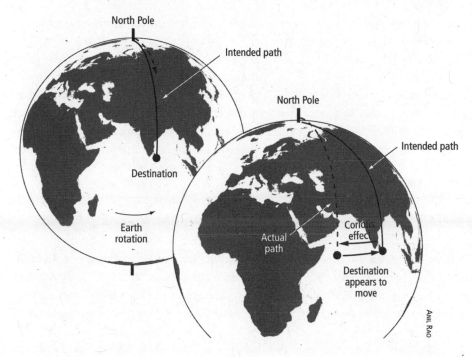

Fig. 2.5: *The Coriolis Effect. The rotation of the Earth causes an apparent deflection in the path of winds. This can be understood by observing the flight of a plane that begins at the North Pole and heads directly south toward Sri Lanka. The plane appears to veer off course. It hasn't. The Earth's rotation makes it look that way.*

that you board a plane leaving the North Pole. The pilot plots a course that will take you due south toward an airfield on Sri Lanka just south of India. If the pilot flies the plane due south the entire trip, will you arrive at your destination in Sri Lanka?

No.

The plane will end up somewhere over the Arabian Sea.

Why?

Because of the Earth's rotation.

As shown in Figure 2.5, as the plane travels south, the Earth rotates beneath it. The Earth rotates eastward. If you plot the flight path of the plane it appears to have been deflected.

It hasn't.

It only looks that way.

The apparent deflection of the plane's path is the Coriolis effect, named after the scientist who first described it. In the Northern Hemisphere, the deflection is to the right of the direction of travel. In the Southern Hemisphere, the deflection is to the left.

Winds flowing north or south also appear to be deflected thanks to the Coriolis Effect. The trade winds, for instance, appear to flow from northeast to southwest.

In the temperate zone, as shown in Figure 2.4c, the low-level north-flowing winds that sweep across the surface of the Earth flow across the North American continent not from south to north but from the southwest to northeast. These are the prevailing southwesterly winds that blow across the Great Plains of

North America. Many a wind farm and many a small wind operation depend on them for electricity. The direction they flow is the result of the Coriolis Effect.

Global airflows provide a source of reliable energy that could be available to use for five billion years as long as their source, the sun, remains. But there's more to wind.

Wind from Storms

As you have just seen, wind flows across the surface of the planet are complicated. They consist of local winds like onshore and offshore winds and enormous global air movements. However, winds are often associated with storms. Storms, in turn, are produced when high-pressure and low-pressure air masses collide. Let's take a brief (and highly simplified) look at the complex interactions that produce fierce, high-energy winds that can be a blessing and a curse to wind energy systems.

As anyone who has watched the evening news realizes, high- and low- pressure zones move across the continents, interacting with one another and creating a wide assortment of weather. Where do these high- and low-pressure air masses come from?

As you might suspect, low-pressure zones originate in the tropics.

Encircling the Earth at the equator is a permanent, elongated, narrow band of low pressure. It's referred to as the equatorial low-pressure trough.

The equatorial low-pressure trough is produced by the huge influx of solar energy in equatorial regions. This heats water and land

masses, which, in turn, warms the air above them, causing it to expand and rise. The atmospheric pressure in the equatorial region is therefore lower than in the Northern or Southern Hemispheres. The result of the constant influx of solar energy is a more or less permanent low-pressure band or trough along the equator. The equatorial low-pressure region is not stable, however. Huge masses of low-pressure air frequently "break off" from the trough and migrate northward, sweeping across the North American continent.

In contrast, the North and South Poles are regions of more or less permanent cold high-pressure air. (Remember cold air is denser than warm air masses and has a higher atmospheric pressure.) Like warm tropical air, huge masses of cold Arctic air also break loose and drift southward, sweeping across the Northern Hemisphere. In the summer, cold Arctic air masses may bring welcome relief to people beleaguered by blisteringly hot temperatures. In the winter, these high-pressure air masses deliver cold Arctic air as far south as Florida and southern California, freezing pipes and sending people and animals scurrying for shelter.

High-pressure and low-pressure air masses, often measuring 500 to 1,000 miles in diameter, move across continents in a magnificent global ballet. As high- and low-pressure air masses collide with one another, they produce an assortment of exciting, sometime dangerous weather. For example, when cold, dry air masses moving in from the north collide with warm, humid air masses from the south, snow and rain form. (Cold air causes the moisture in the warm air masses to condense, resulting in precipitation.)

The movement of high- and low-pressure air masses across continents is steered by prevailing winds and by the jet stream (high altitude winds). Storms produced when these air masses meet are typically characterized by winds of varying intensity. As with local winds and global winds, winds that accompany storms flow between zones of high- and low-pressure. The greater the difference in pressure between a high-pressure air mass and a "neighboring" low-pressure air mass, the stronger the winds. In some cases, these winds contain an enormous amount of energy that can be a blessing to those who rely on wind turbines to meet some or all of their electrical needs. They can also, as noted above, be a curse. Really fierce winds can damage wind systems.

It is often assumed that wind turbines will capture all the energy of a storm. Unfortunately, that's not true. Storms are too infrequent and of too short duration to make it worthwhile to engineer and build a wind generator strong enough to keep spinning and producing electricity in very strong winds. Moreover, a wind turbine engineered to capture the energy of strong storms would not function very well in the low-wind speeds, which are most common.

As you will learn in Chapter 5, all wind generators worth owning have a governing system that reduces the machine's exposure to high winds. In small-scale generators, machines start shedding energy in the 25 to 30 miles per hour range.

Friction and Turbulence and Smart Siting

Now that you understand how winds are created, it's time to fine-tune your knowledge of wind. We'll begin by looking at a major force in the life of wind. It's called ground drag and occurs when air flows across a surface. Ground drag is caused by friction when air flows across a surface.

Friction is the force that resists movement of one material against another. You create friction, for example, when you rub your hands together. (Try it.) When wind flows across land or water, friction occurs.

Ground drag due to friction varies considerably, depending on the texture or roughness of the surface. The rougher or more irregular the surface, the greater the friction. As a result, air flowing across the surface of a lake generates

Fig. 2.6: *Effect of Ground Drag. Winds move more slowly at ground level due to ground drag, and wind speed increases with height above ground, due to a reduction in ground drag.*

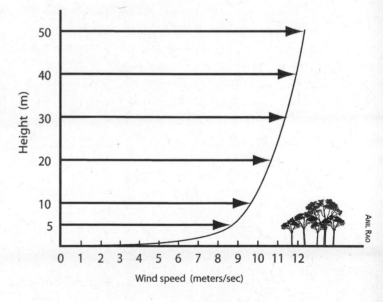

Height (m)

Wind speed (meters/sec)

ANIL RAO

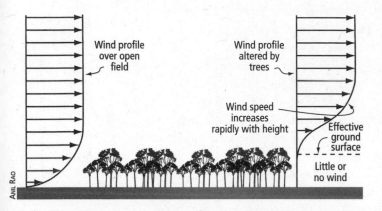

Fig. 2.7: *Wind Speed vs. Height. These graphs compare the wind speed over a grassy area and a forest. As you can see, a forest virtually eliminates ground-level winds, causing the effective ground level to shift upward. Note also that the wind speed increases more rapidly with height over a forest than over a grassy area. Wind turbines placed well above the tree line can avail themselves of powerful winds.*

less friction than air flowing over a meadow. Air flowing over a meadow generates less friction than air flowing over a forest.

Ground drag slows air movement and extends to a height of about 1,650 feet (500 meters) above the Earth's surface. However, the greatest effects are closest to the ground — the first 60 feet above the ground over a relatively flat, smooth surface. Over trees, the greatest effects occur within the first 60 feet (18 meters) above the tree line.

A 20 mile per hour wind measured at 1,000 feet above the surface of the ground covered with grasses, flows at 5 miles per hour at 10 feet. It then increases progressively until it breaks loose from the influence of the ground drag or friction. Figure 2.6 shows the difference

in wind speed at 165 feet (50 meters) to 16.5 feet (5 meters). Figure 2.7 compares wind speed over a grassy area to wind speed over a forest, a significantly rougher surface. Notice that the wind speed increases more rapidly above the forest.

Understanding ground drag is important because of the dramatic way it influences wind speed near the surface of the ground where residential wind generators are, by necessity, located. Because the effects of friction decrease with height above the surface of the Earth, savvy installers typically mount their wind machines on towers 80 to 120 feet high (24 to 37 meters), sometimes as high as 180 feet (55 meters) in forested regions, so their turbines are out of the influence of ground drag. At these heights, the winds are substantially stronger than they are near the ground. As we will make clear shortly, a small increase in wind speed can result in a substantial increase in the amount of power that's available from the wind and the amount of electricity a wind generator can produce. Mounting a wind turbine on a tall tower therefore maximizes the electrical output of the machine. Placing a turbine on a short tower has just the opposite effect. It places the generator in the weaker winds and is a bit like mounting solar panels in the shade.

Ground drag isn't the only natural phenomenon that affects the production of wind machines. Another is turbulence.

Turbulence results from air flow across rough surfaces. Objects in the way of the wind, such as trees or buildings, interrupt the wind's smooth laminar flow, causing it to tumble and swirl (Figure 2.8). Rapid changes in wind speed

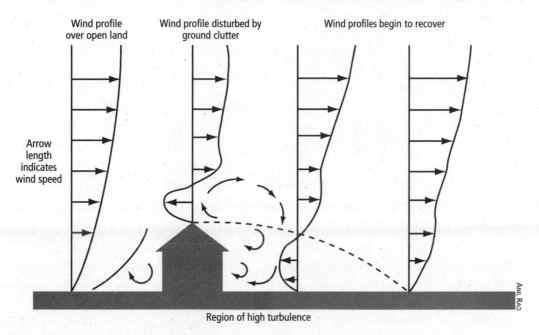

Fig. 2.8: *Ground Clutter and Turbulence. Trees, houses, barns, silos, billboards, garages and other structures — collectively referred to as ground clutter — create considerable amount of turbulence. Like eddies behind rocks in streams, the turbulent zone contains a fluid (air) that swirls and tumbles, moving in many directions. Turbulence reduces the harness-able energy of the wind and causes more wear and tear on wind machines, damaging them over time.*

occur behind large obstacles and winds may even flow in the opposite direction to the general airflow. This highly disorganized wind flow is referred to as turbulence.

Turbulent wind flows wreck havoc on wind machines, especially the less expensive lighter-weight wind turbines often installed on short towers by cost-conscious homeowners in an effort to save money. Buffeted by turbulent winds, wind machines hunt around on the top of their towers, constantly seeking the strongest wind, starting and stopping repeatedly. This decreases the amount of electricity a turbine generates.

Turbulence not only decreases the harvestable power available to a wind generator, it also causes vibration and unequal loading (unequal forces) on the wind turbine, especially the blades, that may weaken and damage the machine. Turbulent winds created by obstacles in the path of the wind therefore increase wear and tear on wind generators. Over time, they can destroy a turbine. The cheaper the turbine, the more likely it will be destroyed, if placed in a turbulent location. A homeowner may find that a machine he or she had hoped would produce electricity for ten to twenty years only lasted two years. In

Obstacles slow the wind down, reducing its harvestable power, but also create cross currents that are difficult for wind turbines to capture.

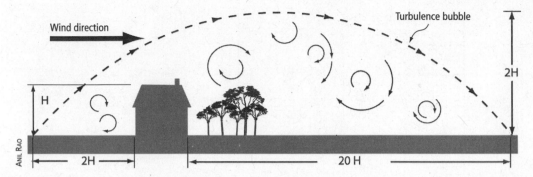

Wind direction

Turbulence bubble

2H

H

2H

20 H

Fig. 2.9: *Turbulence Bubble. The turbulence bubble created by an obstacle or several obstacles like houses and surrounding trees, extends upwind, downwind and even above the clutter.*

Turbulence is to a wind machine what potholes are to your car.

— Robert Preus, Abundant Renewable Energy

extreme cases, an inexpensive, lightweight turbine may be destroyed in a matter of months.

When considering a location to mount a wind turbine, be sure to consider turbulence-generating obstacles such as silos, trees, barns, houses and even other wind turbines, a topic we will discuss in more detail in Chapter 4. Proper location is the key to avoiding the damaging effects of turbulence. Turbulence can also be minimized by mounting a wind turbine on a tall tower.

Mounting a wind generator higher and out of turbulent winds offers four distinct benefits: (1) it situates the wind generator in the stronger higher-energy-yielding winds, substantially increasing electrical production; (2) it distances the machine from the damaging effects of turbulence; (3) it decreases the wind turbine's maintenance and repair requirements; and (4) it increases the wind turbine's useful lifespan substantially, perhaps tenfold. Longer turbine life means less overall expense — and more electricity from your investment.

As shown in Figure 2.9, all obstacles create a downstream zone of turbulent air, or "turbulence bubble." As illustrated, the turbulence bubble typically extends vertically about twice the height of the obstruction. It also extends downwind approximately 15 to 20 times the height of the obstruction. For example, a 20-foot-high house creates a turbulence bubble that extends 40 feet above the ground and 300 to 400 feet downwind from it. As illustrated in Figure 2.9, the turbulence bubble also extends upwind a bit — about two times the object's height. In this case, the upwind bubble extends about 40 feet upwind from the house. The upstream portion of the bubble is created by wind backing up as it strikes the obstacle — much like water flowing against a rock in a river or against the bow of a boat.

If 50-foot trees surrounded the house, the upwind bubble would extend about 100 feet. The bubble would extend downwind about 750 to 1,000 feet and vertically about 100 feet above the ground. Remember also that the bubble shifts as wind direction shifts. If the wind is blowing out of the north, the bubble extends to the south. If the wind is blowing from the south, the bubble extends to the north.

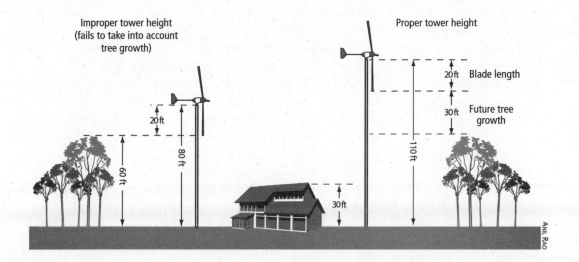

Improper tower height
(fails to take into account
tree growth)

Proper tower height

20ft Blade length

30ft Future tree
growth

20 ft

80 ft

60 ft

110 ft

30ft

ANIL RAO

To avoid costly mistakes, installers recommend that wind machines be mounted so that the complete rotor (the hub and the blades) of the wind generator is *at least* 30 feet (9 meters) above the closest obstacle within 500 feet (about 150 meters), or a tree line in the area, whichever is higher (Figure 2.10). This should place a wind machine out of the turbulence bubble into the stronger, smooth-flowing winds. Don't listen to those who recommend lesser heights. It's advice that you'll regret forever, as many unhappy customers have.

Raising a wind machine into the smoothest, strongest winds ensures greater electrical production and longer machine life — both money-making benefits of smart siting. If your home or business is in an open field surrounded by trees, as is often the case in some Midwestern and Eastern states, remember that trees can dramatically reduce the speed of the wind flowing across your property. The wind turbine needs to be above the tree line to be effective. Remember, too, when determining the appropriate tower height: trees grow and towers don't. Be sure to account for growth of nearby trees over the 20- to 30-year life span of your wind system when determining tower height. (We'll talk more about siting a wind turbine in Chapter 4.)

The Mathematics of Wind Power

We've been emphatic about the importance of placing a wind machine on a high tower above the slower and more turbulent ground winds and will remind you of this essential law of wind energy production a few more times in this book.

To understand how important tall towers are, you need to understand a simple mathematical equation to calculate the power available to a wind turbine from the wind. This equation takes into account the three main factors that influence the output of a wind energy system: (1) air density, (2) swept area, and (3) wind speed.

Fig. 2.10: *Siting a wind turbine in an open area surrounded by trees is possible, but requires special attention to tower height. In this case, the top of the tree line is the effective ground level. The turbine needs to be mounted well above the trees to achieve optimum output.*

The power available to a wind machine is expressed in the equation: $P = \frac{1}{2} d \times A \times V^3$.

In this equation, P stands for the power available in the wind (not the power a wind generator will extract — that's influenced by efficiency and other factors). Density of the air is d. Swept area is A. Wind speed is V. Of the three factors that influence power (P), air density is the least important.

Air Density

Air density is the weight of air per unit volume. Air density varies with elevation, but doesn't have much of an effect on power until one reaches 2,500 feet (about 760 meters) above sea level. As shown in Table 2.1, at 3,000 feet (about 910 meters) above sea level, the air density is only 9 percent lower than at sea level. At 5,000 feet (about 1,525 meters) above sea level, air density has dropped by about 15 percent. At 8,000 feet (about 2,440 meters) above sea level, air density plummets by about 24 percent. You can anticipate a decrease in the expected output of a wind machine of about 3 percent per 1,000 feet (about 300 meters) increase in elevation.

In general, then, for higher elevation areas like Dan's home in the foothills of the Rockies at 8,000 feet above sea level, the air is considerably less dense than lower elevation air in Wisconsin where Mick resides or coastal Washington where Ian lives. A wind turbine in a 30-mile-per-hour wind in Wisconsin or Washington will produce more electricity than an identical wind machine at Dan's residence, assuming they are all mounted at the same height and are free from obstructions. If Mick, Dan, and Ian, for instance, each installed an ARE110, a wind turbine produced by Oregon-based company, Abundant Renewable Energy, and the turbines were operating in a 12-mile-per-hour average wind, Ian's turbine would produce, on average, about 420 kilowatt-hours of electricity per month. (See sidebar for a definition of kilowatt-hours.) Mick at about 700 feet elevation would produce slightly less, approximately 410 kilowatt-hours per month. Dan's machine would produce 318 kilowatt-hours per month.

Table 2.1 Air Density vs. Wind Turbine Performance		
Elevation (ft)	**Elevation (m)**	**Air Density**
0 (sea level)	0 (sea level)	100%
500	152	99%
1,000	305	97%
2,000	610	94%
3,000	915	91%
4,000	1,220	88%
5,000	1,524	85%
6,000	1,829	82%
7,000	2,134	79%
8,000	2,439	76%
9,000	2,744	73%
10,000	3,049	70%

AMERICAN WIND ENERGY ASSOCIATION

Air density is also a function of humidity — the amount of moisture contained in air. Interestingly, humid air is less dense than drier air at the same temperature.[1] As a result, a wind turbine in a 30-mile-per-hour wind in Arizona will produce slightly more electricity than would the same generator in Florida. (The difference, however, is negligible.)

Temperature also affects the density of air. As you would expect from previous material in this chapter, warmer air is less dense than colder air. Consequently, a wind turbine operating in cold (denser) winter winds blowing at 30 miles per hour would produce slightly more electricity than the same wind machine in warmer winds blowing at the same speed. As an example, notes Mick, you get 13 percent more energy from a 30-mile-per-hour wind at 0°F during the winter in a cold climate than at 60°F in the spring due to increased density caused by the difference in temperature.

Although temperature and humidity affect air density, they are not factors we can change. Installers do need to be aware of the reduced energy available at higher altitudes, however, so they don't create unrealistic expectations of a wind-electric system installed in such locations.

Although density is not a factor we can control, wind installers do have control over a couple of other key factors, notably, swept area (A) and, believe it or not, wind speed. As you shall soon see, both of these factors have a much greater impact on the amount of power available to a wind turbine and the electrical output of the machine than elevation, humidity or temperature.

Power and Energy

As you learned in Chapter 1, one electrical measurement used to describe wind turbines is the rated power, measured in watts or, more typically, thousands of watts. One thousand watts is equal to a kilowatt.

A watt is a unit of power and is not to be confused with energy. To the engineer or physicist, energy is the capacity to do work. Power, on the other hand, is the speed at which work is done. That is, power is work per some unit of time. Watts and kilowatts, therefore, represent the rate at which electricity is produced by a wind turbine or consumed in a household. Watt hours or kilowatt-hours, also introduced in Chapter 1, represent the quantity of electricity produced or used. A wind turbine producing electricity *at a rate of* 1,000 watts for one hour produces 1,000 watt-hours or one kilowatt-hour of energy. Energy is what you buy from you electric utility, and it is measured in kilowatt-hours.

Swept Area

Swept area, a topic we'll discuss in greater depth in subsequent chapters, is the area of the circle the blades of a wind machine create when spinning (Figure 2.12). We like to think of it as a wind machine's collector surface. The larger the swept area, the more energy a wind turbine can capture from the wind.

Swept area is determined by blade length.[2] The longer the blades, the greater the swept area. The greater the swept area, the greater the electrical output of a turbine.

As the equation suggests, the relationship between swept area and power output is linear.

Theoretically, a ten percent increase in swept area will result in a ten percent increase in electrical production. Doubling the swept area doubles the output.

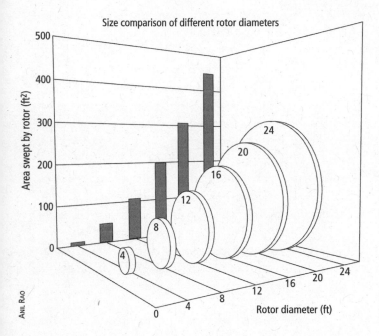

Size comparison of different rotor diameters

Area swept by rotor (ft²)

Rotor diameter (ft)

ANIL RAO

Fig. 2.12: *Blade Length and Swept Area. A small increase in radius or blade length results in a large increase in swept area.*

Shopping Tip

Because swept area is such an important determinant of the output of a wind turbine, we strongly recommend focusing more on the swept area of a wind turbine than on its rated power — at least until the industry can come up with a standardized way of measuring and reporting rated power.[3]

It's important to note, however, that even though the swept area increases the energy output in a linear fashion, relatively small increases in blade length result in very large increases in the swept area.

When shopping for a wind turbine, always convert blade length to swept area, if the manufacturer has not done so for you (they usually do). Swept area can be calculated using the equation $A = \pi \cdot r^2$.

In this equation, A is the area of the circle, the swept area of the wind turbine. The funny-looking Greek symbol is pi, which is a constant: 3.14. The letter *r* stands for the radius of a circle, the distance from the center of the circle to its outer edge. For a wind turbine, radius is the same as the length of each blade.

Because swept area is a function of the radius squared, a small increase in radius or blade length, results in a large increase in swept area (Figure 2-11). As an example, a wind generator with an 8-foot blade has a swept area of 200 square feet. A wind generator with a 25 percent longer blade, that is, a 10-foot blade, has a 314 square-foot swept area. Thus, a 25 percent increase in blade length results in a 57 percent increase in swept area and, theoretically, a 57 percent increase in electrical production.

Wind Speed

Although swept area is more important than the density of air, wind speed is even more important when it comes to the output of a wind turbine. That's because the power available from the wind increases with the cube of wind speed. This relationship is expressed in the power equation as V^3 ($V \times V \times V$), or wind speed multiplied by itself three times. Let's consider a simple example provided by Mick.

Suppose that you mount a wind machine (foolishly, we might add), at 18 feet (5.5 meters) above the ground surface on the grass-covered plains of Kansas. Let's suppose that you measure the output when the wind is blowing at eight miles per hour (3.6 meters per second). A savvy friend, who knows how important it is to mount a wind machine on a tall tower, installs an identical wind turbine on a 90-foot (27 meter) tower. He also measures the output. When the wind is blowing at 8 miles per hour at 18 feet where your wind turbine flies, it is blowing at 10 miles per hour at 90 feet, the height of your neighbor's machine. Wind speed is 25 percent higher.

In this example, the power available to both turbines can be approximated by multiplying the wind speeds by themselves three times. (Units aren't important for this comparision.) For the lower turbine the result is 512 ($8 \times 8 \times 8 = 512$). The power available to the wind turbine mounted on a 90-foot tower is 10 cubed or $10 \times 10 \times 10$ or 1,000.

This example shows that a two-mile-per-hour increase in wind speed, only a 25 percent increase, doubles the available power. Put another way, a 25 percent increase in wind speed yields an increase of nearly 100 percent. As Paul Gipe points out in his book, *Wind Energy: Renewable Energy for Home, Farm, and Business*, "That's why there's such a fuss concerning the proper siting of a wind machine."

Although winds are out of the control of us mortals, homeowners can affect the wind speed at their wind machines by choosing the best possible site and by installing their wind generators on the tallest towers. Remember: the important lesson of V^3 is that a small increase in wind speed results in a very large increase in the power available to a wind turbine and the electrical output of the machine.

Remember for future discussions with those who seem to miss this essential point: a 10 percent increase in wind speed will always result in a 33 percent increase in power available in the wind. That's the reason why we'll remind you repeatedly throughout this book about the importance of proper siting — situating wind machines out of the turbulence bubble and mounting them as high as possible. That's why you should always be suspicious of anyone who tells you that it's fine to mount a wind machine on a short tower in your backyard. It's the worst possible advice. We've heard it given by professional installers and wind turbine manufacturers who, in our mind, are more interested in selling wind turbines than the performance of those wind machines and customer satisfaction.

Closing Thoughts

Wind is a tremendous resource available in many parts of the world thanks to the sun, or, more specifically, the sun's unequal heating of the Earth's surface. Depending on your location, you may be able to take advantage of offshore and onshore winds or perhaps mountain-valley winds. Prevailing winds and winds that form between high-pressure and low-pressure zones could become your ally in achieving energy independence and reducing your carbon footprint. Just don't forget to take into account ground drag created by friction that could slow wind speed near your site and turbulence that can rob you of additional energy and tear your beloved wind machine to pieces. Site wisely and you'll be repaid day after day after day.

CHAPTER 3

WIND ENERGY SYSTEMS

Wind electric systems fall into three categories: (1) grid-connected, (2) grid-connected with battery back up, and (3) off-grid. In this chapter, we'll examine each system. We'll describe their components and discuss the pros and cons of each one. We'll also examine ways to couple a wind system with other renewable energy technologies such as solar electricity to create a hybrid system. This information will help you decide which system, if any, suits your needs and lifestyle. Before we examine the types of wind energy systems, let's take a brief look at their main components.

Components of a Wind Energy System

Wind systems contain many components. To begin, though, we'll focus our attention on two of them: the wind turbine and the tower.

Subsequent chapters contain more detailed discussions of these and other components.

Wind Turbines

The wind turbine or wind generator is a key component of all wind-electric systems. Although there are several types of wind generators, most wind turbines in use today are horizontal axis units or HAWTs (this term is explained shortly). Most small wind turbines in use today have three blades attached to a central hub. Together the blades and the hub are called the rotor. The rotor, in turn, is often connected to a shaft which is directly coupled to an electrical generator. The shaft runs horizontal to the ground, hence the name, horizontal axis wind turbine. When the rotor turns, the generator produces alternating current (AC) electricity. (See the accompanying sidebar for an explanation of AC electricity.)

One of the key components of a successful wind generator is the blades. They capture the wind's kinetic energy and convert it into mechanical energy (rotation) that is converted into electrical energy by the generator. We'll discuss their design, how they work, and the materials they are made of in Chapter 5.

The generators of wind turbines are often protected from the elements by a durable housing made from fiberglass or aluminum (Figure 3.1a). However, in many modern small wind turbines, the generators are exposed to the elements (Figure 3.1b). As Ian points out, the alternator can deal with the weather and benefit from the natural cooling that occurs when it is exposed directly to the air.

Wind generators produce electricity when the wind blows — but only above a certain minimum speed. This is known as the cut-in speed.

Most wind turbines in use today have tails (aka tail vanes) that keep them pointed into the wind to ensure maximum production. However, some very successful turbines like those made by the Scottish company Proven (pronounced pro-vin, not pru-vin) are designed to orient themselves to the wind without tails. (More on these turbines in Chapter 5.)

AC vs. DC Electricity

Electricity comes in two basic forms: direct current and alternating current. Direct current electricity consists of electrons that flow only in one direction through the electrical circuit. DC electricity is the kind produced by flashlight batteries or the batteries in cell phones, laptop computers, or portable devices such as iPods. It is also the kind of electricity produced by photovoltaic modules.

Most wind turbines produce alternating current electricity. In alternating current, the electrons flow back and forth.[1] That is, they switch or alternate direction in very rapid succession, hence the name "alternating current." Each change in the direction of flow (from left to right and back again) is called a cycle.

In North America, electric utilities produce electricity that cycles back and forth 60 times per second. It's referred to as 60-cycle-per-second — or 60 hertz (Hz) — AC. The unit hertz commemorates Heinrich Hertz, the German physicist whose research on electromagnetic radiation served as a foundation for radio, television and wireless transmission. In Europe and Asia, the utilities produce 50-cycle-per-second AC.

Alternating current electricity is produced by most wind generators, but also by electrical generators in hydroelectric and power plants that run on fossil fuels or nuclear fuels. No matter what form of energy is used to turn a generator, all of them operate on the principle of magnetic induction — they move a wire through a magnetic field (or vice versa) to produce electricity. This is true for both AC and DC generators.

a

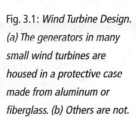

Fig. 3.1: *Wind Turbine Design. (a) The generators in many small wind turbines are housed in a protective case made from aluminum or fiberglass. (b) Others are not.*

b

Towers

Much like the star quarterback on a football team, the wind turbine typically receives most of the attention in discussions on small wind energy. However, like football, wind energy is a team effort. Several other components are parts of the team. They play key roles in capturing and converting the wind's energy, roles that help the star of the team, the wind turbine, shine.

One vital team member is the tower. Although we'll take a much closer look at towers in Chapter 6, a few words are in order now. To begin, residential wind generator towers fall into three broad categories: (a) free-standing, (b) guyed, and (c) tilt up (Figure 3.2).

Freestanding towers may be either monopoles or lattice structures. Freestanding monopole towers consist of high-strength hollow tubular steel like those that support streetlights. Lattice towers consist of tubular steel pipe or flat-metal steel bolted or welded together to form a lattice structure like the famed Eiffel Tower in Paris.

To be successful, freestanding towers must be strong enough to support the weight of the wind turbine, and, more importantly, they need to be strong enough to withstand the forces of the wind acting on the turbine. (Any tower strong enough to do this is strong enough to support the weight of the turbine.) Towers must also have a large foundation to counteract

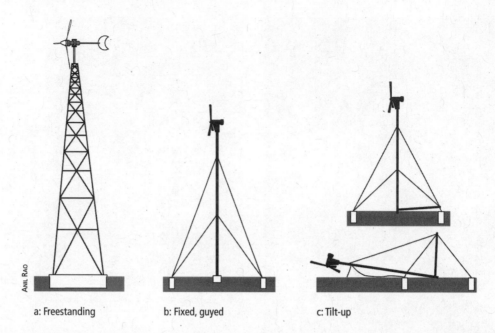

ANIL RAO

a: Freestanding   b: Fixed, guyed   c: Tilt-up

Fig. 3.2:
Tower Types
Three types of
towers in use today:
(a) freestanding,
(b) fixed, guyed and
(c) tilt-up.

the tremendous forces applied by the wind, forces that could easily topple a weak or poorly anchored tower. Large amounts of steel and concrete are required to accomplish this task. This makes freestanding towers the most expensive tower option.

The second tower type, the guyed tower, is shown in Figure 3.2. Guyed towers may be made of high-strength steel pipe or may be lattice structures, and are supported by high-strength steel cables, known as guy cables. The guy cables extend from anchors buried securely in the ground to tower attachment points. Guy cables stiffen towers so they don't buckle, and they hold them in place so they don't lean or fall over.

Because guy cables strengthen and stabilize the tower, less steel is required for a guyed

tower than a freestanding tower.[2] As illustrated in Figure 3.2, guy cables are attached to the tower every 20 to 30 feet, although wider spacing is also possible. A 120-foot-tall tower may require four sets of cables.

The third type of tower, and one that is very popular, is the tilt-up tower, so named because it can be raised and lowered — tilted up and down. This feature makes it possible to inspect, maintain and repair a wind turbine from the ground. Tilt-up towers typically consist of high-strength steel pipe or a lattice structure supported by guy cables.

When it comes to the performance of a wind system, towers are as important as the wind turbine itself. Unfortunately, many wind turbines are mounted on towers that are too short. This occurs out of ignorance and misguided frugality. As we will point out numerous times in this book, installing too short a tower is a foolish, potentially costly mistake. Mounting a wind generator on a short tower is akin to mounting a solar electric module in the shade. As we have pointed out, properly sized towers place wind turbines in the more powerful winds where the energy is. Properly sized towers also raise turbines out of turbulence created by ground clutter that severely diminishes the quality and quantity of wind. Returning to our football analogy, towers are like the offensive line in a football team. They protect the star player, the turbine, making its life less rough and tumble and ensuring much higher productivity. Don't let anyone talk you into a short tower!

Wind systems also require wires that run down the tower, transporting electricity produced by the turbine to the point of use. The wires attach to leads on the generator and typically run externally — for example, alongside a tower leg of a lattice tower — to a junction box at the base of the tower. From there, they typically run underground in conduit to the house, barn, shop or business.

As you shall soon see, wind-electric systems involve a number of additional components. Most wind systems require an inverter and an AC service panel, the breaker box through which electricity from the wind system flows to household circuits. Battery-based wind energy systems require batteries to store surplus electricity and a device known as a charge controller that helps protect the batteries. Wind systems, like other renewable energy systems, also require

disconnects, meters and protective devices such as fuses, circuit breakers and lightning arresters. We'll describe the function of each of these components as we explore the three main types of wind energy system. More detailed discussions of key components, like towers, turbines, inverters and batteries are found in subsequent chapters.

Wind Energy System Options

As noted in the Introduction, wind systems fall into three main categories: (1) grid-connected, (2) grid-connected with batteries, and (3) off-grid. We'll begin with the simplest, the grid-connected system.

Grid-Connected Systems

The simplest of the major wind energy systems is the grid-connected option, shown in Figure 3.3. Grid-connected systems are so named because they are connected directly to the electrical grid — the vast network of electric wires that criss-cross cities, towns, states and nations. They're also typically referred to as batteryless grid-connected or batteryless utility-tied systems because they do not employ batteries to store surplus electricity.[3]

In batteryless grid-tied systems, the electrical grid accepts surplus electricity — electricity produced by the turbine in excess of household or business demand. When a wind system is inactive, the grid supplies electricity to the home or business. In a sense, then, the grid serves as the storage medium.

As shown in Figure 3.3, a batteryless grid-connected system consists of six main

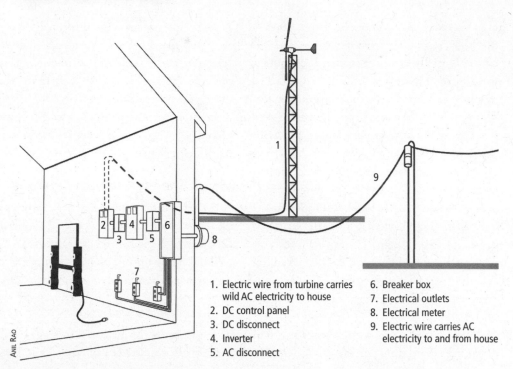

Fig. 3.3: *Grid-Connected Wind System. The grid-connected wind system is the simplest of all systems. Wild AC electricity produced by the turbine is first fed into the controller. The inverter produces grid-compatible AC electricity to power household loads. Surpluses are fed onto the electrical grid.*

ANIL RAO

1. Electric wire from turbine carries wild AC electricity to house
2. DC control panel
3. DC disconnect
4. Inverter
5. AC disconnect
6. Breaker box
7. Electrical outlets
8. Electrical meter
9. Electric wire carries AC electricity to and from house

Frequency and Voltage

Alternating current electricity is characterized by a number of parameters. Two of them are frequency and voltage.

Frequency refers to the number of times electrons switch direction every second and is measured as cycles per second. (One cycle is a switch from flowing to the right to flowing to the left.) In North America the frequency of electricity on the power grid is 60 cycles per second (also known as 60 Hertz). The flow of electrons through an electrical wire is created by an electromotive force that scientists call voltage. The unit of measurement for voltage is volts.

You can think of voltage as electrical pressure, or the driving force that causes electrons to move.

components: (1) a wind generator, specifically designed for grid connection, (2) a tower, (3) an inverter/power conditioner, (4) the main service panel, (5) meters, and (6) the electrical grid.

To understand how a batteryless grid-connected system works, let's begin with the wind generator. In most batteryless grid-connected systems, the wind generator produces "wild AC" electricity.[4] Wild AC is alternating current whose frequency and voltage vary with wind speed. (See sidebar for a definition of frequency and voltage.) In most small wind turbines, the faster the blades spin, the higher the voltage, and the greater the frequency.

Wild AC, produced by wind turbines, is not directly usable. Appliances and electronic devices require a tamer version of electricity — alternating current with a relatively constant frequency and voltage, like that available from the grid. In a grid-connected system, then, the wild voltage must first be "tamed." That is, its frequency and voltage must be converted to standard values.

Wild AC is converted to useful AC in two steps. First, the wind system converts (rectifies) the wild AC to DC. Rectifiers, electronic devices that accomplish this conversion, are typically located in a controller, shown in Figure 3.3. (In some systems, the "controller" rectifiers are in the inverter.) DC electricity flows from the controller to the next component of the system, the inverter, also shown in Figure 3.3. The inverter converts the DC electricity to grid-compatible AC — 60 cycle per second, 120-volt (or 240-volt) electricity. Because the inverter produces electricity in sync with the grid, it's often referred to as a synchronous inverter.

While grid-compatible wind generators typically produce wild AC, another type of wind generator is being introduced into the small wind market that holds great promise. It is known as an induction generator. Explained in more detail in Chapter 5, an induction generator is similar to an electric motor that, when driven fast enough by the spinning blades of the wind turbine, produces grid-compatible AC electricity. (This is the way large commercial wind turbines on wind farms work.) As a result, its output does not need to be converted to DC and then inverted back to AC, eliminating a few costly components of a small wind energy system.

The 120-volt or 240-volt AC produced by the inverter (or directly by an induction generator) flows to the main service panel. From here, it flows to active loads — that is, to electrical devices drawing power. If the wind machine is producing more electricity than is needed, which is often the case; the excess automatically flows onto the grid.

As shown in Figure 3.3, surplus electricity backfed onto the grid travels from the main service panel through the utility's electric meter, typically mounted on the outside of the house. It then flows through the wires that connect the home or business to the grid. The surplus electricity then travels along the power lines where it flows into neighboring homes or businesses. The utility treats the electricity as if it were theirs and the end users pay the utility directly for the electricity you provided to them. Fortunately, the electric meter monitors the home's or business's contribution to the grid and the utility dutifully credits the local producer for its contribution. The meter also keeps track of electricity the power company supplies to the home or business when the wind system is not generating. To learn how the electric company measures what you are putting onto the grid and how they "pay" for it, check out the accompanying box, "Net Metering in Grid-Connected Systems."

In addition to the electric meter — or meters (some utilities require two or more meters) — that monitor the flow of electricity onto and off the local utility grid, code-compliant grid-connected wind energy systems also contain safety disconnects. These are

manually operated switches that enable service personnel to disconnect at a couple of key
points in the system to prevent electrical shock
if service is required. As shown in Figure 3.3, an
AC disconnect is located between the inverter
and main service panel. When switched off, it
disconnects and isolates the wind energy system from the household circuits and the grid.
The AC disconnect must be mounted outside so
that it is accessible to utility company personnel. That way, utility workers can isolate the
wind system from the grid if they need to work
on the electric lines in your area without fear of
shock — for instance, if a line goes down in an
ice storm. The AC disconnect must also contain
a switch that can be locked in the off position
by the utility worker so that the homeowner
or a family member doesn't accidentally turn
the system back on prior to their completion
of repairs. For some time, the lockable disconnect was considered critical for the safety of
utility personnel, because the 120- or 240-V
electricity from an inverter becomes thousands
of volts on the utility distribution line after it
passes through the "service-drop" transformer.

Interestingly, lockable AC disconnects are
not required by all utilities. Moreover, the large
California utilities with thousands of solar-
and wind-electric systems are dropping the
requirements for utility company-accessible,
lockable disconnects. They've found that they
can't use them because there are so many.
Moreover, these companies have gained confidence that they're not needed. As discussed
shortly, grid-compatible inverters automatically
shut off when the utility power goes down. As

Ian notes, "Properly installed wind-electric systems will not backfeed a dead grid, with or
without a disconnect. Period." That's because
grid-compatible inverters are designed so that
they automatically disconnect from the grid
when a power outage occurs. As we'll discuss in
Chapter 8, grid-compatible inverters monitor
line voltage and frequency (the frequency and
voltage of electricity on the grid). When they
detect a change in either, for instance, a drop in
voltage due to a power outage, the inverter automatically shuts down — and stays off until the
situation is corrected. As a result, no electricity can flow onto the grid.

"Lockable disconnects are theoretically
unnecessary, because UL-listed inverters meet
interconnection safety standards and can be
relied upon to disconnect if the grid goes
down," notes Robert Aram. "But when human
life is at stake, many believe that prudence dic-
tates that you not rely on only one system for
safety." He adds, "The lockable disconnect is a
backup to the safety features of the inverter. It
is also on the outside of the building, where
the utility worker can see it — and see that it
is open (disconnected). Utility workers can
also put their personal locks on them to be
sure that it stays open (disconnected)." Robert
continues, "The inverter is inside the building
where the utility worker cannot see it. How
does he know that you have a UL-listed
inverter? Perhaps you have kluged something
together that feeds directly into the grid with
no safety functions?" He concludes, "Every
source of power in a utility's system, every generator and every transformer, has a visible

disconnect switch. Since the utilities have backup safety disconnects for all their own equipment, isn't it reasonable for them to require renewable energy system owners to have one also?"

The Pros and Cons of Grid-Connected Systems

Batteryless grid-connected systems represent the majority of all new wind systems in the United States. Although popular, they do have their pluses and minuses, summarized in Table 3.1.

On the positive side, batteryless grid-connected systems are relatively simple. And, as such, they are generally less expensive than other options. They are often 25 percent cheaper than battery-based systems.

Because batteryless grid-connected systems contain no batteries, they require less maintenance. Operators won't have to manage and maintain a costly battery bank, a topic discussed in Chapter 7. Nor will the owner have to replace costly batteries every five to ten years. Moreover, no batteries means no battery box or battery room, which may be costly to add to an existing home or incorporate into a new home design. All of these factors add up to potentially huge savings.

Another substantial advantage of batteryless grid-connected systems is that they can store an unlimited amount of electricity (so long as the grid is operational). Although these systems don't store excess electricity on site like a battery-based system for later use — for example, during periods when the wind's not

blowing — they "store" surplus electricity on the grid in the form of a credit on your utility bill. When winds fail to blow — or a wind machine isn't producing enough electricity to meet demands — electricity can be drawn from the grid, using up the credit. The grid serves as an unlimited battery bank. (See sidebar for a discussion of grid storage.)

Unlike a battery bank, Mick points out in his workshops, you can never "fill up" the grid. It will accept as much electricity as you can feed it. In contrast, when batteries are full, they're full. They can't take on more electricity. Surpluses generated in wind-electric systems not connected to the grid are used to heat a living space or water so as not to be wasted.

Another advantage of grid-connected systems with net metering is that utility customers suffer no losses when they store surplus electricity on the grid. With battery storage, they do. As described in detail in Chapter 7, when electricity is stored in a battery, it is converted to chemical energy. When electricity is needed, the chemical energy is converted back to electrical energy. As much as 20 to 30 percent of the electrical energy fed into a battery bank is lost due to conversion inefficiencies.

In sharp contrast, electricity stored on the grid comes back in full. If you deliver 100 kilowatt-hours of electricity, you can draw off 100 kilowatt-hours. (We'd be remiss if we didn't point out the grid has losses too, but net metered customers get 100 percent return on their stored electricity.)

Another notable advantage of batteryless grid-tie systems is that they are greener than

Net Metering in Grid-Connected Systems

Utilities keep track of the two-way flow of electricity from grid-connected renewable energy systems in one of two ways. In many cases, they install a single dial-type electric meter. This meter measures kilowatt-hours of energy and can run forward or backward. It runs forward (counting up) when your home or business is using more energy than your wind turbine is producing. It runs backward (counting down) when the wind turbine is producing more power than is being used.

In many new installations, utility companies install digital net meters that tally electricity delivered to and supplied by a home or business. They don't "run backwards," as older, dial-type meters do. They simply keep track of electricity coming from and going to the grid so the utility can determine whether you've produced as much, more, or less than you've consumed.

Still other utilities install two meters, one to tally the electricity delivered to the grid and another to keep track of electricity supplied by the grid. (As a side note, they typically charge to install the second meter and charge a separate monthly fee to read the second meter.)[5]

If one meter is installed, the company typically offers a billing service called net metering. If two meters are installed, the arrangement is referred to as net billing or "buy-sell." Let's take a look at the net metering option first.

Net metering is a system in which the electric bill is based on net consumption — consumption minus production. That is to say, a customer's electric bill is based on the amount of utility energy consumed minus the amount of energy provided to the grid from a renewable energy system. The "net" in net metering refers to net kilowatt-hours.

In a buy-sell or net billing arrangement, one meter tracks consumption. The other tallies a customer's contribution to the grid. The customer is charged for the electricity he or she consumes and is credited for the electricity he or she contributes to the grid, often at different rates. The "net" in net billing refers to net dollars. Let's look at a few examples to see how these systems work and the difference between them.

Let's start with net metering. Suppose that a wind energy customer delivered 1,000 kilowatt-hours of electricity to the grid during the month of October, a fairly windy month in most locations in North America. Suppose also that the customer consumed 1,000 kilowatt-hours of grid energy during that month. If the utility engaged in net metering, the cost of electricity to the customer for the month would be zero, although utilities typically charge a $5 to $20 per month fee to read the meter and to pay for maintenance of the grid. This fee may be referred to as a "customer service charge" or something similar. Utilities also charge taxes and other incidentals, for example, surcharges to fund pollution abatement or renewable energy rebate programs. In this example, then, the customer would simply pay the monthly fee and applicable surcharges.

Consider another example of net metering, one in which the customer produced more electricity ☞

than consumed. In this instance, let's suppose that the customer delivered 1,000 kilowatt-hours of electricity, but the home or business only used 500 kilowatt-hours of electrical energy from the grid. In this instance, the customer would only be charged the monthly fee. What would happen to the surplus?

In most cases, utilities carry the surplus from one month to the next. Like many cell phone companies, they carry surpluses for up to a year. If, for instance, the customer that produced a surplus of 500 kilowatt-hours in October consumed 500 kilowatt-hours more than his or her wind system delivered to the grid the next month, the customer and the utility are even.

If, however, the customer produced a surplus of 500 kilowatt-hours in October and produced 1,000 kilowatt-hours more than he or she consumed in November and a 500 kilowatt-hours surplus in December, all of these surpluses would be carried forward. At the end of the year, surpluses, if any, are either paid for by the utility or, as is more common, forfeited. This system is known as annual net metering.

In some cases, for example, in Wisconsin where Mick lives or in British Columbia, where Dan has clients, utilities reconcile the bill each month (although in Mick's case, he negotiated an annual net metering agreement). This arrangement is known as monthly net metering. How does it work? Let's consider an example.

Suppose that a customer produced a surplus of 500 kilowatt-hours of electricity in the month of January, typically one of the windiest months of the year. The utility charges 10 cents per kilowatt-hour for electricity supplied by the grid and credits customers the same amount for surpluses they deliver to the system. In this instance, then, the customer would receive a check for $50 minus meter reading and other fees. (That's 500 kilowatt-hours surplus x 10 cents per kilowatt-hour minus meter reading and other fees.) If you're thinking that wind could be a profitable venture, don't get your hopes up. While some utilities write checks for surpluses, many don't. They "take" the surplus without payment to the customer — it all depends on state law. In some states, the utility pays wholesale rates for the surplus, which are usually one-quarter to one-third the retail rate. (If you're not happy with your state law, we suggest that you work to change it!)

As a rule, monthly net metering is generally the least desirable option, especially if surpluses are "donated" to the utility company and not reimbursed or reimbursed at wholesale rates. Annual reconciliation is a much better deal. It permits wintertime surpluses to be "banked" to offset summertime shortfalls, if any. However, don't forget that even with annual net metering, the end-of-the-year surplus, if any, may simply be lost, donated to the utility. If that's the case, Ian points out that the ideal arrangement from a customer's standpoint is to negotiate a continual month-to-month roll-over, so there's no concern about losing credit for your wind-powered kilowatt-hours.

Net metering is mandatory in many states, thanks to forward-thinking, renewable energy ☞

legislators. At this writing (December 2007), 43 states and the District of Columbia have instituted net metering, although there are substantial differences between them, the programs vary with respect to who is eligible, which utilities are required to participate in the program, the size and types of systems that qualify, payment structure, and so on. Many states only require "investor-owned" utilities to offer net metering, but they typically serve urban regions where wind energy systems are generally not a good idea. In many states, municipal power companies and rural cooperatives are exempt from net metering laws. "The exemption of rural co-ops from net metering is a big problem for small wind," notes Jim Green. "Rural areas are more likely to have opportunities for wind due to larger property sizes, less ground clutter to obstruct the wind, and fewer zoning barriers than urban or suburban areas."

Utilities that don't offer net metering may use another system, discussed earlier. It's called buy-sell or net billing. In these buy-sell schemes, utilities typically install two meters, one to track electricity the utility sells to the customer, and another to track electricity the customer feeds onto the grid. Unlike net metering, utilities that engage in buy-sell arrangements typically charge their customers retail rates for electricity they use but pay customers wholesale rates for electricity fed onto the grid by renewable energy systems. For example, a utility may charge 10 to 15 cents per kilowatt-hour for electricity they supply to the customers, but pay wholesale rates of 2 to 3 cents per kilowatt-hour for power that customers deliver to the grid. How does this work out financially for the small-scale producer?

As you might suspect, not very well.

Suppose that a customer consumed 500 kilowatt-hours of electricity from the grid but fed 1,000 kilowatt-hours of surplus electricity onto the grid in December, also a typically windy month. Let's suppose that the retail rate for electricity was ten cents per kilowatt-hour and the generation and delivery costs (that is, the wholesale cost) came to three cents per kilowatt-hour. In this case, the utility would charge the customer 10 cents per kilowatt-hour for electricity delivered to them, or $50 for the 500 kilowatt-hours. The utility would credit the customer 3 cents per kilowatt-hour for the 1,000 kilowatt-hours of electricity delivered to the grid. That is, the customer would be paid or credited $30.00 for the 1,000 kilowatt-hours of electricity sold to the utility. As a result, the customer would end up owing the utility $20 plus a meter reading fee of $10 to $20. The customer comes out, once again, on the bottom of the midden heap and guess who ends up on top?

Had the customer been net metered, he or she would have profited from the deal. The utility would have charged the customer $50 for the electricity purchased from the utility but would have paid or credited them $100 for the 1,000 kilowatt-hours of electricity fed onto the grid, if they paid retail rates for surplus. The customer would ☞

have been $50 ahead of the game, minus meter-reading and other minor fees.

Although (at this writing) seven states do not have net metering laws, progress in this area is moving quickly. More and more utilities have come to realize that it is often cheaper — and less hassle — for them to net meter than to install two meters. As Mike Bergey wrote in his article entitled "A Primer on Small Turbines" posted on the American Wind Energy Association's website, "Because of high overhead costs to the utilities for keeping a few special hand-processed customer accounts, net metering is actually less expensive for them."

For information about net metering rules for individual states and utilities, go to the Database for State Incentives on Renewables and Efficiency (dsireusa.org). To find out if your state requires net metering and to learn the specifics about their program, you can log on to the US Department of Energy's website: eere.energy.gov/state_energy/policy_content.cfm?policyid=26.

To learn more about net metering, log on to *Home Power* magazine's website: homepower.com/resources/net_metering_faq.cfm. The American Wind Energy Association has a great fact sheet on net metering on their website, awea.org. Check out "fact sheets" under "publications." ■

battery-based systems. Although utilities aren't the greenest entities in the world, they are arguably greener than battery-based systems.

Batteries require an enormous amount of energy to produce. Lead must be mined and refined. Batteries must be assembled and shipped to homes and businesses, requiring additional energy. Batteries contain highly toxic sulfuric acid. Although old lead-acid batteries are recycled, they're often recycled under less than ideal conditions. In less developed countries, children are often employed to remove the lead plates along the banks of rivers. Acids from batteries may contaminate surface waters.

Yet another advantage of batteryless grid-connected systems, when net metered, is that they can provide substantial economic benefits. In windy sites, these systems may produce surpluses month after month. If the local utility net meters and pays for surpluses at the end of the month or end of the year, these surpluses can generate income that helps reduce the cost of a wind system and therefore the annual cost of producing electricity.

On the downside, grid-connected systems may require careful and sometimes annoyingly difficult negotiations with local utilities. While many utilities are cooperative, others may throw up roadblocks. Some utilities, especially some of the rural electric associations, are hostile to the idea of buying electricity from customers interested in installing a renewable energy system.

Grid Storage — Is Electricity Stored on the Grid?

Many people are confused when renewable energy experts talk about "storing" electricity on the grid. There's good reason for this.

Technically speaking, renewable energy isn't stored on the electric grid, as it is in batteries. In batteries, electricity is converted to chemical energy for storage. When the electricity is needed, the chemical energy is converted back into electrical energy.

In a grid-connected system, surplus electricity flows onto the grid. It is not physically or chemically stored on the grid, however. Rather, it is immediately consumed by one's nearest neighbors.

That said, electricity fed onto the grid is *effectively* stored thanks to net metering. Here's what we mean: when surplus electricity is back fed onto the grid, a homeowner or business is credited for the surplus. The utility "banks" the electricity. That is, it

sells the electricity to another customer, but keeps track of what has been delivered to the grid, so it can pay the supplier back later.

By keeping track of the surplus electricity produced by a net-metered renewable energy system, the utility says, "You've supplied us with *x* kilowatt-hours of electricity. When you need electricity, for example, when the wind is dead, if we practice net metering, we'll supply you with an equal amount at no cost." In a sense, the utility *has* stored the electricity for its customer. Bear in mind, however, when the utility gives back the electricity you've "stored" on the grid, they supply you with electricity most likely generated by coal or nuclear energy or natural gas. The net effect on your utility bill is the same, however.

(We'll discuss this topic in detail in Chapter 10.)

Another downside of batteryless grid-connected systems, worth serious consideration, is that they are vulnerable to grid failure. That is, when the grid goes down, so does a batteryless grid-connected wind system. Homes and businesses cannot use the output of a batteryless wind or photovoltaic (PV) system when the grid is not operational.

Even when winds are blowing, batteryless grid-tied wind energy systems shut down if the grid experiences a problem — for instance, a line breaks in an ice storm or lightning strikes

a nearby transformer, resulting in a power outage. Even though the winds are blowing, you'll get no power from your system.

If power outages are a recurring problem in your area and you want to avoid service disruptions, but like the idea of being connected to the grid, you may want to consider installing a standby generator that switches on automatically when grid power goes down. Bear in mind, however, that a standby or backup generator takes many seconds to start and come on line, so your power is interrupted during this time. If you want to avoid this temporary interruption, you could install an uninterruptible power supply (UPS) on critical equipment such as computers. An uninterruptible power supply contains a battery and an inverter. If the utility power goes out, the UPS will supply uninterrupted power until its battery get low. Or, you may want to consider installing a grid-connected system with battery backup. In the latter, you'll need to install a battery-based grid-tie inverter and batteries. Batteries provide backup power to a home or business when the grid goes down.

Grid-Connected Systems with Battery Backup

Grid-connected systems with battery backup are also known as battery-based utility-tied systems. These systems ensure a continuous supply of electricity, even when freezing rain wipes out the electrical supply to you and another million utility customers in your part of the country.

As shown in Figure 3.4, a grid-connected system with battery backup contains many of the components found in grid-connected systems: (1) a wind turbine on a tower, (2) a controller, (3) an inverter, (4) safety disconnects, (5) main service panel, and (6) meters to keep track of electricity delivered to and drawn from the grid.

Although grid-connected systems with battery backup are similar to batteryless grid-connected systems, they differ in several notable ways. The most notable difference is that battery-based grid-connected systems require batteries. In addition, battery-based grid-connected systems require a different type of

Table 3.1 Pros and Cons of Batteryless Grid-Tie Systems	
Pros	**Cons**
Simpler than other systems	Vulnerable to grid failure unless an uninterruptible power supply (generator) is installed
Less expensive	
Less maintenance	
More efficient than battery-based systems	
Unlimited storage of surplus electricity	
Greener than battery-based systems	

inverter. (We'll describe the differences in Chapter 8.)

The third major difference is a meter that allows the operator to monitor the flow of electricity into and out of the battery bank, and the fourth is a device known as a charge controller.

Batteries for grid-connected systems with battery backup are either flooded lead-acid batteries or, more commonly low-maintenance sealed lead-acid batteries. However, battery banks in grid-connected systems are typically small. That's because they are usually sized to provide sufficient storage to run a few critical loads for a day or two until the utility company restores electrical service. These vital appliances include a refrigerator, a few lights, and the blower of a gas or oil furnace or the boiler and pump in a gas- or oil-fired radiant floor system. For those who want or need full power during outages, the battery bank must be much larger, and the larger the battery bank, the greater the cost of the system. If you don't want to install a larger battery bank, you can install a generator to meet a large demand for electricity.

Another important point worth noting is that keeping batteries fully charged is a high priority of these systems. Battery banks in grid-connected systems are maintained at full charge day in and day out to ensure a ready supply of electricity should the grid go down. It's only when the batteries are topped off and household demands are being met that excess electricity is back fed onto the grid.

In battery-based grid-tie systems, batteries are called into duty only when the grid goes down. They're a backup source of power. They're not there to supply additional power, for example, to run loads that exceed the wind system's output. When demand exceeds supply, the grid makes up the difference, not the batteries. When the winds are dead, the grid, not the battery bank, becomes a home or business's power source.

Yet another point worth noting is that maintaining a fully charged battery bank requires

Fig. 3.4:
Grid-connected wind system with battery backup

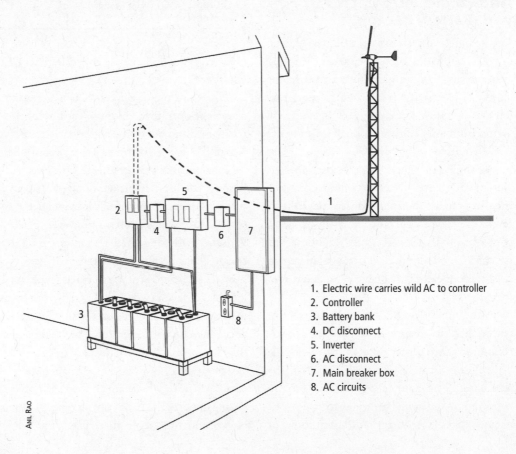

1. Electric wire carries wild AC to controller
2. Controller
3. Battery bank
4. DC disconnect
5. Inverter
6. AC disconnect
7. Main breaker box
8. AC circuits

Anil Rao

a fair amount of electricity. That is to say, a portion of the surplus electricity a wind system generates may be devoted to keeping batteries full.

Batteries require this continual input of electricity because they self-discharge. They lose electricity when sitting idle. You've seen it happen to flashlight batteries or car batteries sitting idle for months. Batteries therefore become a regular load on renewable energy systems.

How much electricity is required to top off a battery bank?

In the best case, Ian notes, topping off batteries will consume five to ten percent of a system's output. In the worst case — that is, in a system with a low-efficiency, unsophisticated inverter that is used to charge a large or older battery bank — it may approach 50 percent.

Battery banks in grid-connected systems don't require careful monitoring like those in off-grid systems, but it is a very good idea to keep a close eye on them. When an ice storm ravages utility lines in your area and knocks out power to your home or business, the last thing you want to discover is that your battery bank is dead.

For this reason, grid-connected systems with battery backup typically include a meter to monitor the total amount of electricity stored in the battery bank. These meters give readings in amp-hours or kilowatt-hours. (See sidebar for definitions.) You'll learn more about meters to monitor batteries in Chapter 7.

Meters also typically display battery voltage. Battery voltage can provide a very general

Amps and Amp-Hours

Electricity is the flow of electrons through a wire. Like water flowing through a hose, electricity flows through conductors at varying rates. The rate of flow depends on the voltage (defined earlier).

The flow of electrons through a conductor is measured in amperes or amps for short. (An amp is 6.23×10^{18} electrons passing by a point on a conductor per second.) The greater the amperage, the faster the electrons are flowing.

One amp of electricity flowing through a wire for an hour is one amp-hour. This term is also frequently used to define a battery's storage capacity. A flooded lead-acid battery, for example, might store 420 amp-hours of electricity.

approximation of the amount of energy in a battery — if you know how to interpret this parameter. We'll discuss this topic in Chapter 7.

Another component found in wind energy systems with battery backup is the charge controller, shown in Figure 3.4. As you may recall from our discussion of batteryless grid-connected systems, the controller in these systems contains a rectifier. It converts AC to DC. The controllers in grid-connected systems with battery backup also contain rectifiers. DC electricity is then fed into the batteries. Charge controllers, in grid-connected systems with battery backup also monitor battery voltage. They use this information to protect batteries from being overcharged — having too much electricity driven into them.

When the charge controller sees that the batteries are fully charged — they have reached the "full" voltage set point and hence

HOMEPOWER.COM

Fig. 3.5: *Dump load resistive heaters like this one are used as dumps for surplus electricity from battery-based wind energy systems.*

are in danger of overcharging — it terminates the flow of electricity to them. Surplus electricity is then fed onto the grid, or if the grid is not operational, to a diversion or dump load — typically a resistance-type device that accepts the surplus electricity and turns it into heat. Resistance heaters are installed in existing water heaters or as separate space heaters in the basement or a nearby utility room and serve as dump load, making some use of the surplus energy (Figure 3.5).

Charge control is essential to battery-based systems because overcharging batteries can permanently damage their lead plates, dramatically reducing battery life. Most wind turbines designed for battery charging come with an appropriate charge controller.

Batteries also require protection from discharging too deeply, referred to by the techies

as overdischarging. Like overcharging, overdischarging damages the lead plates of batteries, dramatically reducing their productive life. To prevent over discharging, charge controllers contain a low-voltage disconnect (LVD), although they typically only cover DC loads — that is, circuits that draw DC electricity directly from the battery bank to supply DC appliances and lights. To protect against overdischarging by AC loads, many battery-based inverters incorporate a low-voltage AC disconnect. It shuts down the inverter AC output if the battery voltage drops too low.

Low-voltage disconnects in the charge controller or inverter terminate the flow of electricity from the battery bank when the amount of electricity stored in batteries falls to 20 percent of the battery's storage capacity.

In battery-based grid-tie systems, deep discharge is a rare event. Overdischarging typically only occurs during extended utility power outages.

Pros and Cons of Grid-Connected Systems with Battery Backup

Grid-connected systems with battery backup protect against utility failures. They enable homeowners to continue to run critical loads — that is, select loads in their homes during power outages, such as energy-efficient refrigerators and energy-efficient lighting while neighbors grope around in the dark and their food thaws in their freezers and begins to rot in their refrigerators. These systems allow businesses to protect computers and other vital electronic equipment so they can continue operations while their competitors twiddle their thumbs and complain about losses.

Although battery backup may seem like a desirable feature, it does have some drawbacks. For one, grid-connected systems with battery backup cost more to install and operate. Flooded lead-acid batteries and sealed batteries used in these and other renewable energy systems are expensive. Remember, too, that a battery bank needs a home. If you are building a new home or office, you need will to add a well-ventilated battery box or battery room that stays warm in the winter and cool in the summer to house your batteries. In addition, unless you install a very large battery bank, your home or business won't be protected against blackouts lasting more than three or four days. Nor will you be able to meet all of the power requirements of your home or business.

Flooded lead-acid batteries also require periodic maintenance and replacement. As explained more fully in Chapter 7, to maintain batteries for long life, you'll need to monitor their fluid levels regularly and fill them with distilled water every few months. Battery banks also need to be vented to the outside to prevent potentially dangerous hydrogen gas buildup, which can lead to explosions and fires if ignited by a spark. In addition, battery banks need to be replaced periodically, whether you use them or not. Typical batteries used in this system require replacement every five to ten years, at a cost of several thousand dollars each time.

Yet another problem is that battery based grid-tie systems require a substantial portion of the daily renewable energy production just to keep the batteries fully charged.

Because grid-connected systems with battery backup are expensive and infrequently required, few people install them. For about ten years, Mick had a huge battery bank in the cellar of his wind-powered home in Wisconsin. This large and costly battery bank was called into service only a couple of times because power outages in his area are infrequent and short-lived, rarely lasting longer than ten to twenty minutes. In ten years, his family experienced few power outages that lasted longer than a few hours. Moreover, Mick freely admits that he didn't perform the required maintenance, so when he needed the system, the batteries didn't perform well. They'd lost much of their storage capacity and therefore provided only a fraction of the electricity his family needed. He admits that this was a case of "out of sight, out of mind." Because of his experience, he rarely recommends battery backup to clients.

When contemplating a battery-based grid-tie system, Ian points out that you need to ask yourself three questions: (1) How frequently does the grid fail in your utility's service area? (2) What are your critical loads and how important is it to keep them running? (3) How do you react when the grid fails?

If the local grid is extremely reliable, you don't have medical support equipment to run or need computers for critical financial transactions, and don't mind using candles on the rare occasions when the grid goes down, why buy, maintain and replace costly batteries?

Table 3.2
Pros and Cons of a
Battery-Based Grid-Tie System

Pros	Cons
Provides a reliable source of electricity	More costly than batteryless grid-connected systems
	Less efficient than batteryless grid-connected systems
	Less environmentally friendly than batteryless systems
	Requires more maintenance than batteryless grid-connected systems

In some cases, people are willing to pay for the reliability that a battery bank brings to a grid-connected system. One of Dan's clients in British Columbia buys and sells stocks, bonds and currency for hedge fund clients. He can't experience down time during active trading, not for a second. As a result, he opted for a grid-connected system with battery backup for his home and office.

See Table 3.2 for a quick summary of the pros and cons of battery-based grid-connected systems.

Off-Grid (Stand-Alone) Systems

Those who want to or must supply all of their needs through wind energy or a combination of wind and some other renewable source need to install an off-grid or stand-alone system. As shown in the schematic drawing of an off-grid system in Figure 3.6, this system bears a remarkable resemblance to a grid-connected system with battery backup. There are some

noteworthy differences however. Before we examine them, let's take a look at how an off-grid system operates.

As illustrated in Figure 3.6, the main source of electrical energy in an off-grid system is a battery-charging wind turbine. These turbines typically produce 12-, 24- or 48-volt DC electricity, although some battery-charging generators like the Bergey Excel produce 120- and 240-volt DC electricity.

Although battery-charging wind generators produce DC electricity, we should point out that they actually produce wild AC electricity first. It is converted (rectified) to DC electricity. The rectifiers may be located in the generator or in the controller.

Electricity flows from the turbine to the controller. The controller delivers DC electricity to the battery bank in this system. When electricity is needed in a home or business, it is drawn from the battery bank via the inverter. The inverter converts the DC electricity from the battery bank, typically 24 or 48 volts in a standard system, to higher-voltage AC, either 120 or 240 volts, which is required by households and businesses. The AC then flows to active circuits in the house via the main service panel.

Although off-grid systems resemble grid-connected systems with battery banks, there are some noticeable differences. The first and most obvious is lack of grid connection. That is, there are no power lines running from the house or business to the grid. These systems stand alone. They are self-sufficient power systems in which the wind turbine produces all of the electricity required to meet the needs of a family or business. Off-grid systems rely on batteries to store surplus electrical energy generated during windy periods for use during low- or no-wind periods.

A second major difference between an off-grid system and a grid-connected system with battery backup is that the off-grid system often requires assistance, in the form of a PV array, a micro hydro turbine, or a gasoline or diesel generator, often referred to as a gen-set. One or more of these energy sources helps make up for shortfalls.

Although backup generators are commonly used in off-grid renewable energy systems, Mick contends that properly sized wind/PV hybrid systems rarely, if ever, require them. In fact, he's retrofitted numerous PV systems with wind generators to avoid the need for generator backup. Ian point outs, however, that the majority of off-grid systems include generators. He notes that it takes a very balanced mix of solar and wind resources to avoid the need for a gen-set. "A gen-set also provides redundancy," notes Jim Green. "If a critical component of a hybrid system goes down temporarily, the gen-set can fill in while repairs are made."

Like grid-connected systems with battery backup, an off-grid system requires safety disconnects to permit safe servicing. A DC disconnect is located between the wind turbine

1. Carries wild AC from turbine
2. Controller
3. Battery bank
4. Transfer switch
5. Inverter
6. Subpanel (for critical loads)
7. Main breaker box (all household loads)
8. Utility meter
9. Service line

ANIL RAO

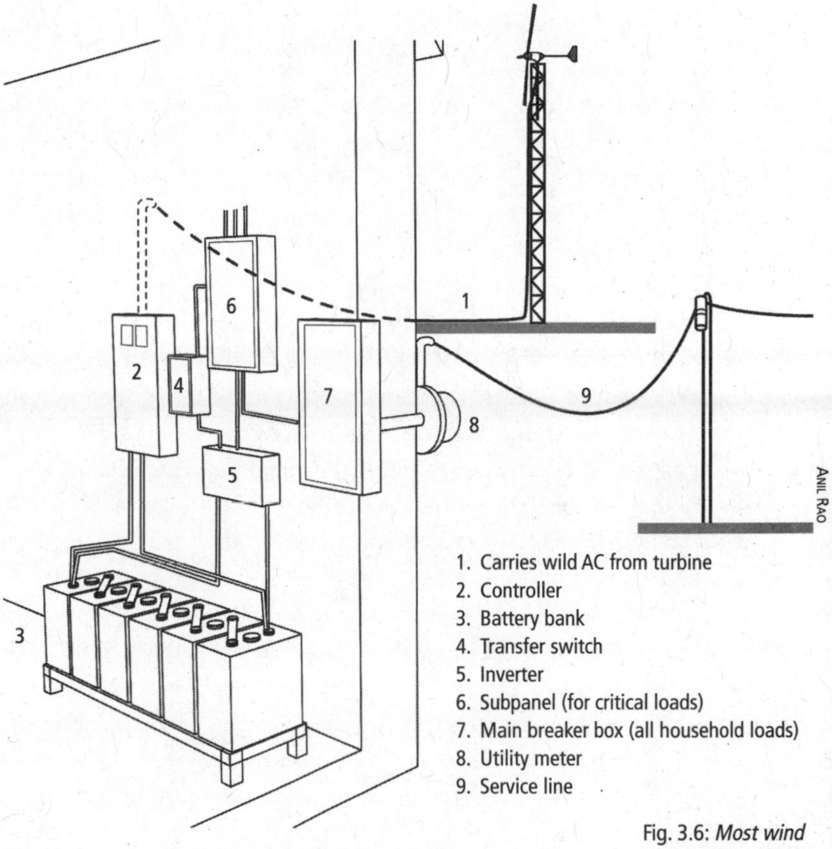

Fig. 3.6: *Most wind turbines in off-grid wind systems produce AC electricity that's converted to DC electricity by the controller. The inverter draws electricity from the batteries, converting DC electricity into AC electricity for household consumption.*

and charge controller. These systems also contain charge controllers to protect the batteries from overcharging. Off-grid systems also typically incorporate a low-voltage disconnect to prevent deep discharge of the battery bank.

As is evident by comparing schematics of the three types of systems, off-grid wind energy systems are the most complex. Moreover, some systems are partially wired for DC — they contain DC circuits that are fed directly from the battery bank to service lights or DC refrigerators.

DC Circuits in an Otherwise AC World?

Most modern homes and businesses operate on alternating current electricity. However, off-grid homes supplied by wind or solar electricity — or a combination of the two — can be wired to operate partially or entirely on direct current electricity. DC-powered lights, refrigerators, televisions and even ceiling fans must be installed in such instances.

One reason to wire a home for DC-only operation is that DC systems do not require inverters. Electricity flows directly out of the battery bank to service loads. This, in turn, may reduce the cost of the system because small household-sized inverters cost $2,500 to $4,000.

Unfortunately, cost savings created by avoiding an inverter may be offset by other factors. For one, appliances designed to operate on DC electricity cost considerably more than their AC counterparts. DC ceiling fans, for instance, cost four times more than comparable AC models. You could pay $200–$250 for a DC model, but only $40–$50 for a comparable AC ceiling fan.

DC appliances and electronics are not only more expensive, they are more difficult to find. You won't find them at Best Buy or other local appliance and electronics retailers.

Many DC appliances are tiny, too. Many DC refrigerators, for example, are miniscule compared to the AC models used in homes. That's because DC appliances are designed for boats and recreational vehicles, and there's not a lot of room in either one for large appliances.

For these and other reasons, DC-only systems are rare. They are typically installed only in remote cabins and cottages.

If you are thinking about installing an off-grid system in a home or business, your best bet is an AC system, although you may want to consider installing a few DC circuits. Dan's off-grid home, powered by solar electricity and wind, uses AC electricity almost entirely. However, when wiring the home, he included a DC circuit to power a DC pump that pumps water from his cistern to the house. He also included a DC circuit to power three DC ceiling fans, although he bought AC fans and runs AC electricity to them because of their lower cost.

Another reason for avoiding use of an inverter is efficiency. As you will see in Chapter 8, most inverters for off-grid systems are about 90 to 95 percent efficient. That is, they consume some energy — 5 to 10 percent — when converting DC to AC and boosting the voltage. Bypassing the inverter with a DC circuit to the water pump or refrigerator reduces this loss and can result in large savings over the long haul.

Even though modern refrigerators are much more energy efficient than their predecessors, they still are major energy consumers in modern homes. Refrigeration often accounts for 10 to 25 percent of a family's daily electrical demand. Because refrigerators consume so much electricity, a DC unit like those offered by Sun Frost or Sun Danzer, can save a substantial amount of energy over the long run (Figure 3.7). Those thinking about an off-grid ☞

system may want to consider DC refrigerators and DC freezers like the Sun Danzer chest freezer.

It is important to note that, while efficient, DC refrigerators and freezers do not come with the features that many Americans expect, such as automatic defrost. They also cost much more than standard or even high-efficiency AC units. Because of these reasons, Ian rarely recommends the inclusion of DC circuits for off-grid installations. They're only for a certain type of renewable energy user — people who are trying to wring every possible kilowatt-hour from a small system.

There are other reasons to think twice about DC circuits in an otherwise AC home. For one, low-voltage DC circuits must be wired with larger gauge wire and connections should be soldered. DC circuits also require special plugs and sockets which are more expensive than their AC counterparts. Richard Perez, founder of *Home Power* magazine, argues that DC appliances are typically not as reliable or well built as their AC counterparts. They are primarily designed for intermittent use in recreational vehicles and tend to wear out more quickly and require more frequent replacement than AC appliances. If that's not enough to dissuade you from DC appliances, many AC appliances have no DC counterparts.

Arguing in favor of AC installation, Perez notes, "The main advantage of using AC appliances is standardization. The wiring is standard — inexpensive, conventional house wiring. The appliances are standard, and are available with a wide variety of features." Furthermore, he adds, "The appliances are designed for regular use, and most are reliable and well built."

Perez points out that while "the main disadvantage of using AC appliances in off-grid systems is the cost of the inverter and the energy lost due ☞

SunFrost

Fig. 3.7: *SunFrost Refrigerator.*
Arguably the most energy-efficient refrigerator on the market, the SunFrost refrigerator is a durable, hard-working unit. Many modern refrigerators, however, are now nearly as efficient, and have many more features customers want.

to inversion inefficiency," modern inverters have an average operating efficiency similar to the amount of energy lost in low-voltage DC wiring, especially if the wire runs are long or are not terminated properly. Therefore, DC may not save any energy at all by bypassing the inverter.

Still, you may want to consider installing a DC circuit in the utility room — or wherever the inverter is located — in off-grid or grid-connected systems with battery backup, to power a DC compact fluorescent light bulb in case of emergency. That way, if the system goes down at night, and you need to find out why, you'll have some light. Be sure to have a spare DC light bulb on hand, too. ■

Fig. 3.8: *Power centers like this one contain all of the components needed for a successful installation, all mounted on one panel. They're easy to wire and pass inspection with ease.*

(For a discussion of DC circuits and DC appliances, see the accompanying box.)

To simplify installation of battery-based systems, many installers recommend use of a power center, such as the one shown in Figure 3.8. Power centers contain many of the essential components of a renewable energy system, including one or more inverters, and the meters needed to monitor system performance. They also contain safety disconnects and charge controllers. Power centers also provide busses (connection points) to which the wires leading to the battery bank, the inverter and the wind generator connect. Although power

WIND ENERGY SYSTEMS 69

centers may cost a bit more than buying all the components separately, they are easier and cheaper to install.

Power centers also make the lives of electrical inspectors much easier and make inspectors very happy. (A happy inspector is a good thing!) While some electrical inspectors know a fair amount about renewable energy systems, many haven't had much experience in this area. When one of the less experienced inspectors shows up on a job site and encounters a system equipped with a power center that's approved by Underwriter's Lab (UL), he or she can be assured that all of the vital components are present and accounted for. Inspections are a snap.

Pros and Cons of Off-Grid Systems

Off-grid systems offer many benefits, including total emancipation from the electric utility (Table 3.3). They provide a high degree of energy independence that many people long for. As Mick points out, while you may be independent from the utility, you are not totally independent. You'll need to buy a gen-set and fuel, both from large corporations. Gen-sets produce pollution and cost money to maintain and operate. If designed and operated correctly, off-grid systems will provide energy day in and day out for many years. Off-grid systems also provide freedom from occasional power failures (although the cost of this protection may be high, as Mick discovered.)

Off-grid systems do have some downsides. As you might suspect, they are the most expensive of the three options. Battery banks and

Table 3.3 Pros and Cons of Off-Grid Systems	
Pros	**Cons**
Provide a reliable source of electricity	Generally the most costly wind energy system
Provide freedom from the utility grid	Less efficient than battery-less grid-connected systems
Can be cheaper to install than grid-connected systems if located more than 0.2 miles from grid	Require more maintenance than batteryless grid-connected systems (you take on all of the utility's operation and maintenance jobs and costs)

generators add substantially to the cost. They also require more wiring. You will also need space to house battery banks and generators. Dan had to build an insulated, soundproof ventilated shed to house his backup generator to appease neighbors who complained about the noise. As noted earlier, batteries also require periodic maintenance and replacement every five to ten years, depending on the quality of batteries you buy and how well you maintain and treat them.

Although cost is a major downside, there are times when off-grid systems cost the same or less than grid-connected systems — for example, if a home or business is located more than a few tenths of a mile from the electric grid. Under such circumstances, it can cost more to run electric lines to a home than to install a small off-grid system. (We'll talk more about this in later chapters.)

To Stand Alone or Not to Stand Alone?

Going off the grid may sound like a glorious way to live your life — free of utility bills and connection to the grid. As we've noted in the text, however, the off-grid option comes at a price. When Mick consults with individuals who want to go off the grid for philosophical reasons, he cautions them to consider the ramifications of this decision, especially if they are philosophically opposed to utilities and concerned about the environmental impacts of grid-generated electricity. He does not at all condone how utilities operate, but from a purely environmental perspective, he thinks it is extremely difficult to justify a battery bank and gen-set system over grid connection.

One reason for this is efficiency. The grid delivers coal-generated electricity at about 30–33 percent efficiency. A Honda gen-set charging a battery bank operates at about five percent efficiency, Mick notes.

The emissions from that coal-fired plant are regulated by the EPA — more or less. Although there are also emission regulations for small engines (manufactured since September 1, 1997), the emissions per kilowatt-hour of electricity are far greater from the backyard generator than the coal-fired power plant. In fact, if everyone had a gen-set in their back yard to meet their electrical demands, we'd all suffocate. Visit Bejing or any number of cities in developing countries where generators are used to produce electricity and you'll see the effect.

If you are thinking about going off the grid, the responsible approach might be to install a hybrid system in which wind is supplemented by some other clean, renewable energy technology, such as PV or micro hydro.

When installing an off-grid system, remember that *you* become the local power company and your independence comes at a personal cost. It also comes at a cost to the environment, because the production of lead-acid batteries is not benign. As noted earlier, although virtually all lead-acid batteries are recycled, battery production is responsible for considerable environmental degradation. Mining and refining the lead, for instance, are fairly damaging. Lead production and battery recycling are often carried out in poor countries with lax or nonexistent environmental policies. According to Mick, they are responsible for some of the most egregious pollution and health problems facing poorer nations. So, think carefully before you decide to install an off-grid system. For more on this topic, see the accompanying sidebar.

Hybrid Systems

As pointed out in this chapter, individuals and businesses have three basic options when it comes to wind-energy systems. Each of these systems can be designed to include two or more renewable energy sources (Figure 3.9).

Hybrid renewable energy systems are extremely popular among homeowners in rural areas. In fact, most residential off-grid wind systems in use in North America are hybrids that combine solar electricity with wind. Moreover, one of the most active segments of the residential wind energy market is the PV-only system owners who are expanding their systems to incorporate wind energy.

Wind and PV are a marriage made in heaven in many parts of the world. In many locations in North America, like the Great Plains, winds are consistent and strong throughout the year. Consider Greensburg, Kansas, the small town destroyed by a tornado in the spring of 2007. Table 3.4 presents average monthly wind speeds in Greensburg at 120 feet above the flat short-grass prairie. As you can see, Greensburg experiences robust winds year round.

In most other locations, winds vary throughout the year. They tend to be strongest in the fall, winter, and early spring — from October or November through March or April. Table 3.5, shows the average wind speed at Dan's educational center in east central Missouri. As you can see, the strongest winds occur from October through May. During these months,

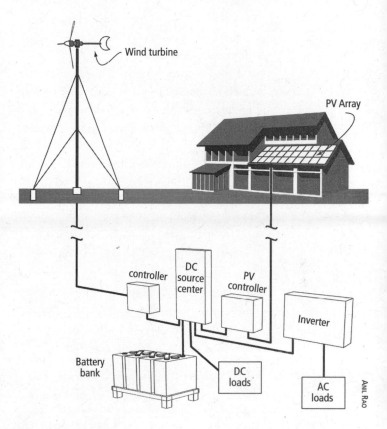

Fig. 3.9: *Hybrid System.*
To meet their needs, many homeowners and business owners install hybrid systems consisting of two or more energy sources, often wind and solar electricity.

Table 3.4													
Ten-Year Monthly Average Wind Speed in Greensburg, Kansas, at 120 Feet													
Month	**Jan**	**Feb**	**Mar**	**Apr**	**May**	**Jun**	**Jul**	**Aug**	**Sep**	**Oct**	**Nov**	**Dec**	**Annual Average**
10-year Average													
meters/second	6.55	6.63	7.50	7.60	6.97	6.60	6.26	6.01	6.30	6.36	6.43	6.44	6.64
miles/hour	14.65	14.83	16.80	17.00	15.60	14.76	14.00	13.44	14.09	14.23	14.38	14.41	14.85

a properly sized wind generator can meet most of a family or business's needs. June through September is less windy.

Winds continue to blow, but less frequently and less forcefully. Fortunately, sunshine is more abundant during this period, as shown in Table 3.6. Installing a solar electric system would supplement the wind-generated electricity. It could provide the bulk of the electricity while the wind turbine played a subsidiary role, backing up the PV system. Together, they can supply 100 percent of the annual electrical demand. This complementary relationship is shown graphically in Figure 3.10.

Because solar and wind resources are often complementary, hybrid systems provide a more consistent year-round output than either wind-only or PV-only systems.

Wind and solar energy often complement each other and can be combined to create a reliable, year-round source of electricity. Note how well solar makes up for the reduction in wind during the summer months at this site.

Sized correctly in areas with a sufficient solar and wind resource, a hybrid wind/PV system can not only provide 100 percent of one's electricity, it may eliminate the need for a backup generator. What is more, hybrid systems often require smaller solar electric arrays and smaller wind generators than if either were the sole source of electricity.

If the combined solar and wind resource is not sufficient throughout the year or the system is undersized, a hybrid system may require a backup generator. A gen-set will supply electricity during periods of low wind and low

Table 3.5
Ten-Year Monthly Average Wind Speed Near Gerald, Missouri, at 120 feet

Lat 38.325 Long 91.297	Jan	Feb	Mar	Apr	May	Jun	Jul	Aug	Sep	Oct	Nov	Dec	Annual Average
10-year Average													
meters/second	5.96	5.93	6.50	6.31	5.27	4.83	4.40	4.31	4.64	5.16	5.67	5.82	5.40
miles/hour	13.33	13.26	14.54	14.12	11.79	10.8	9.84	9.64	10.38	11.54	12.68	13.02	12.08

NASA SURFACE METEOROLOGY

Table 3.6
Solar Resource near Gerald, Missouri, measured in kWh/m² per day*

Lat 38.325 Long 91.297	Jan	Feb	Mar	Apr	May	Jun	Jul	Aug	Sep	Oct	Nov	Dec	Annual Average
Tilt 38	3.66	3.86	4.71	5.43	5.28	5.50	5.75	5.66	5.48	4.79	3.27	2.95	4.70

*Note: This data represents solar energy striking a collector mounted at an optimal angle for this location.

NASA SURFACE METEOROLOGY

sunshine. It is also used to maintain batteries in peak condition, as discussed in Chapter 7, and it may allow installation of a smaller battery bank.

Choosing a Wind Energy System

Although you have a lot of choices when it comes to installing a wind system, by far the cheapest and simplest option is a batteryless grid-connected system.

Although you may encounter an occasional power outage, which shuts your system down, in most places these are rare and transient events.

Grid-connected systems are cheaper than other options. If you are building a new home, however, and you are more than a a few tenths of a mile from existing power lines, connecting to the grid could be expensive. Although some utility companies foot the bill for line extension, you may be charged to run an electrical line to your home. Be sure to check, when considering which system you should install. "Utility policies vary considerably when it comes to line extension costs," notes Jim Green. Hook-up fees can be upwards of $50,000 if you live half a mile from the closest electric lines. Rates vary, so be sure to check. "I have a client in central Indiana who is building seven-eighths of a mile from the existing power line," notes Robert Aram. "His local utility quoted $19,000 to run the line to his building site."

If you are installing a wind system on an existing home that is already connected to the grid, it's generally best to stay connected. Use the grid as your battery bank. When installing such a system, you'll need to sign an intercon-

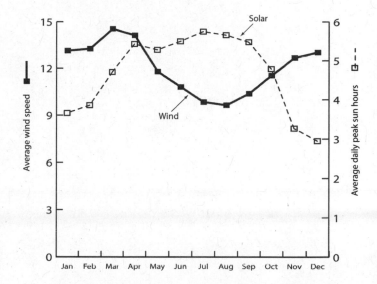

nection agreement with the utility company, and work out a billing agreement — hopefully annual net metering that credits you at retail rates for surpluses.

Grid-connected systems with battery backups are suitable for those who want to stay connected to the grid, but also want to protect themselves from occasional blackouts. They cost more, but provide peace of mind and real security.

Although more expensive than grid-connected systems, off-grid systems are often the system of choice for people in remote rural locations. When building a new home in a rural location, it may be cheaper to install an off-grid system than a grid-connected system.

You don't have to live that far from the grid for an off-grid system to make sense. In some locations, a quarter-mile grid connection is costly enough to justify an off-grid system.

Fig. 3.10: *The complementary nature of wind and solar.*

CHAPTER 4

WIND SITE ASSESSMENT

Wind is an abundant and clean energy resource available in many locations throughout the world for commercial and residential use. Because it is renewable, this vast resource will be available to humankind for as long as its source — the sun — continues to warm our planet. Unfortunately, wind energy systems are not suitable for all locations. So before you invest your hard-earned money in a wind system, it is important to determine whether wind makes sense for you.

For many people, making sense means economic sense. We should point out, however, that a wind energy system could make sense to others for environmental reasons. Or perhaps installing and maintaining a wind energy system might be a fun and interesting hobby, one that pays a little back. Dan has clients for whom his wind site and economic assessments indicated that a wind system would provide a

return on investment well under one percent per year but who decided to install the system anyway. They want a wind turbine and are willing to put their money where their values are. They made a commitment to the future of their children and grandchildren and the rest of humanity.

As a rule, the decision to install a wind energy system is based on a combination of factors. We talked about some of them in Chapter 1 when we discussed the benefits of wind turbines. We'll discuss some of them again in this chapter, which looks at costs in detail.

If the economics of a wind system is your primary or perhaps only focus, the decision to invest time and money on a wind system hinges on three key factors: (1) your electrical energy requirements — how many kilowatt-hours of electricity you need, (2) the number of kilowatt-hours of electricity a wind turbine

The Economics of Wind

Several factors help to make small wind economically viable and can make it a more profitable venture. They include: (1) high electricity rates, (2) rebates or tax credits from utilities or governments, (3) a good wind resource, and (4) a long-term perspective.

In California, with very generous rebates (about 40 percent of system cost), extremely high electrical rates — ranging from 20 cents to 40 cents per kilowatt-hour — small wind power makes perfect sense. In Hawaii, where electricity costs 25 cents per kilowatt-hour, wind makes sense in any good wind site even *without* incentives. In Germany, where electricity costs 75 cents per kilowatt-hour, wind is an even better investment. On the other hand, in some parts of Washington state, where hydroelectricity costs a paltry three to four cents per kilowatt-hour, it's nearly impossible for wind to compete economically. The same can be said for British Columbia, where electricity costs just over six cents per kilowatt-hour.

could produce at your site, and (3) the cost of supplying electricity from a wind system versus some other source, for example, the local utility, or perhaps some other renewable energy technologies, such as solar electricity or micro hydro. If a wind system produces electricity at a rate that's competitive with conventional power or other renewable sources you are considering, it may be a good decision. If it can't compete economically, it may not be a good financial investment. Even if money isn't a primary driving force, most of us want to know if a wind energy system — or a hybrid wind energy system — represents an intelligent investment. This chapter will help you analyze the investment decision.

To begin, let's explore how you determine electrical demand. This will tell you how much electricity you'll need to generate from a wind-electric system. Although we'll focus principally on residential demand, this discussion also pertains to a host of other applications, among them cabins, cottages, workshops, offices, farms and ranches.

Assessing Electrical Demand

Electrical energy consumption varies widely from one household to the next. In an all-electric home equipped with a wide assortment of electric appliances — especially energy-consuming electric stoves, central air conditioners, electric space heating, and electric hot water — average monthly electrical consumption typically falls within the 2,000 to 3,000 kilowatt-hour range. In homes that use natural gas or propane to cook, heat water, and provide space heat — and have no air conditioner — electrical consumption may be as low as 400 to 500 kilowatt-hours per month. On average, though, most homes in the United States consume between 800 to 1,000 kilowatt-hours of electricity per month.

How one goes about assessing the electrical consumption of a residence or business depends on whether it is an existing structure or one that's about to be built. Let's begin with existing structures.

Assessing Demand in Existing Structures

Assessing the electrical consumption of an existing home or business is a fairly easy task. Simply review monthly electric bills, going back two to three years, if possible. (In some areas, every utility bill includes a year-to-date summary of electrical consumption.)

If you don't save electric bills, don't be dismayed. A telephone call to your utility will usually suffice. Most utilities are happy to provide data on monthly and annual energy use to their customers. In addition, many companies allow customers to look up their utility bills online. All you need is your customer number.

If you recently purchased an existing home, you can ask the previous owner for their utility bills. If they have not saved them, they may be willing to contact the utility company to request a summary of electrical consumption over the past two to three years. Remember, however, a house doesn't consume electricity, its occupants do, and we all use energy differently. If you are frugal and a previous owner and his family were not, their past energy usage may be of little value to you in predicting your future usage.

To determine total annual electrical consumption from past utility bills, simply add up the monthly bills. If you have bills for two or more years, you can calculate a yearly average

> ## Average Annual Electrical Use
>
> Based on national averages, homes consume about 1,000 watts continuously, or about 24 kilowatt-hours per day. That's nearly 12,000 kilowatt-hours per year or about $1,200 worth of electricity a year (not counting meter reading fees, taxes and surcharges) at 10 cents per kilowatt-hour.

from this data. If several years' data are available, it is helpful to determine the range — that is, the lowest year and the highest year.

Once you have calculated total annual electrical consumption in kilowatt-hours and determined the range, you next calculate monthly averages — that is, how much electricity is used, on average, during each month of the year. To do this, simply add up household electrical consumption for each month, and then divide by the number of years' worth of data you have. For example, if your records go back four years, add the electrical consumption for all four Januarys, and divide by four. Do the same for February and each of the remaining months. Finally, record monthly averages on a table. This is especially useful to those interested in off-grid systems, because you will want to size the system to meet your needs during the months of highest electrical demand. Individuals interested in installing grid-connected systems generally only need annual electrical consumption data.

After you have calculated annual energy consumption, the range and the monthly averages, look for patterns or trends in energy use. Is energy use on the rise, staying constant, or declining from year to year? If you notice a dramatic increase in electrical energy demand in recent years, you may want to exclude earlier data, which lowers the average. More recent data more closely reflects your electrical consumption. If, on the other hand, electrical energy consumption has declined, earlier data will inflate electrical demand.

It's also a good idea to determine annual cycles — that is, how electrical energy use varies during the year. Are there months or seasons during which electrical demand is higher or lower than others? If your home is in a hot, humid climate, for example, electrical consumption may peak in the summer when the air conditioner is operating full tilt. If you live in a colder climate in which air conditioning is not required, electrical consumption may peak in the winter because lights are on the longest and pumps or blowers of heating systems are working overtime to keep you warm. Understanding seasonal demand is extremely important when sizing a wind system, especially an off-grid system.

Assessing Demand in New Structures

Determining monthly electrical consumption is pretty straightforward when retrofitting an existing home or business with wind or a wind-solar electric hybrid system. When considering a wind energy system to power a new building that's just been built or is about to be built, the task is more difficult because there's no historical use data. What do you do?

One way to estimate future electrical consumption in such instances is to base it on usage in your current home. If you are building a new home that's the same size as your current home and has the same amenities, for example, electrical consumption may be similar to that in your current home. However, if you are building a more energy-efficient home, installing much more energy-efficient lighting and appliances and incorporating passive solar heating and cooling (all of which we highly recommend), electrical consumption could easily be 50 percent, perhaps 75 percent lower than in your current home. If that's the case, adjust current electrical consumption to reflect the savings you'll enjoy.

Another technique that some individuals use is a load analysis. (Load refers to any device that uses electricity.) To perform a load analysis, list all the appliances, lights and electronic devices you expect to install in your new home or office. Then determine how much electricity each one uses when operating and how many hours per day they operate. From this data, you can approximate demand in kilowatt-hours.

Electrical load analysis is pretty straightforward. It's made even simpler by worksheets, like the one shown in Table 4.1. Similar worksheets can be found online, for example, at the Solar Living Institute's website. Professional renewable energy installers also often supply customers with worksheets, and may even help customers fill them out.

When using a chart to estimate electrical consumption, begin by listing all the appliances, lights, and electronic devices in your new home or business. Rather than list every light bulb separately, however, you may want to lump them together. Once you've completed the list, you need to determine the amount of electricity each item consumes. This information can be obtained in a couple of ways. The first is by consulting a chart like the one in Table 4.2. It

Table 4.1 **Electrical Consumption Chart**												
Individual Loads	Qty	X	Volts	X	Amps	=	Watts AC	DC	X Use Hrs/day	X Use days/wk	+ 7 days	= Watts Hours AC	DC
											7	
											7	
											7	
											7	
											7	
											7	
											7	
											7	
											7	
											7	
											7	
											7	
											7	
											7	
											7	
											7	
											7	

AC Total Connected Watts: _____   AC Average Daily Load: _____

DC Total Connected Watts: _____   DC Average Daily Load: _____

lists wattages for a wide range of electrical devices, including stoves, microwaves and stereos. You can find more detailed charts in various books, such as Solar Energy International's *Photovoltaics: Design and Installation Manual* or John Schaeffer's *Solar Living Source Book*. You can also go online for assistance. The website of one Wisconsin utility, WE Energies, has a chart that will help you determine the approximate electrical consumption of appliances and electronic devices. You can find this chart by logging on to webapps.we-energies.com/appliances/apply_calc.cfm.

Energy consumption in a new home can be more accurately determined, however, by consulting the product label or nameplate on the back or bottom of an appliance or electronic device. It typically lists the unit's power consumption in watts. It also typically lists amperage and voltage. Unfortunately, there's no universal standard for reporting this important information. Voltage may be listed as 120 volts, 120 V, 120 volts AC, or 120 VAC. They are all the same. In some cases, however, the stickers only list voltage and current (amperage). To calculate watts, simply multiply the two (watts = amps x volts). This will provide an accurate value of watts for devices that are purely resistive, such as electric heaters and incandescent light bulbs. For devices that run by electromagnetic induction, such as motors and fluorescent light fixtures, amps x volts will overestimate watts. Multiply by a factor of about 0.7 to get a better estimate of watts for these devices.

Load analysis requires use of the "run wattage," that is, the maximum wattage an appliance draws when in operation. However, the run wattage of an appliance typically represents a "worst case estimate," according to Scott Russell, author of a superb article on load analysis, *Starting Smart: Calculating Your Energy Appetite*, in *Home Power* magazine (Issue 102). Russell goes on to say, "Since you rarely watch your television at full volume or use your jigsaw to cut granite, feel free to reduce this number by about 25 percent for 'variable wattage' items such as these." Record this number on the table. In addition, when determining the run wattage of laptop computers and other electronic devices that plug into a charger or transformer, use the information on the charger, not the device itself. This includes cordless drills, cordless phones, and electronic keyboards. Be sure to note DC appliances, if any.

Another way to determine wattage is to use a meter like one of those shown in Figure 4.1. They are plugged into electrical outlets and electronic devices and appliances you want to monitor are then plugged directly into them via a receptacle on the meter. The meter indicates the instantaneous power (watts).

After you have listed all of the electrical devices you'll be using and have recorded the run wattage of each one, you need to estimate the number of hours each device is used each day, which will be discussed shortly. After estimating the amount of time each device is used on a daily basis, you need to determine the number of days each device is used during a typical week in your home or business. From this information, you can calculate the weekly energy consumption of all devices — lights,

Table 4.2
Average Electrical Consumption of Common Appliances

General household
Air conditioner (1 ton) 1500
Alarm/Security system 3
Blow dryer...............................1000
Ceiling fan...............................10-50
Central vacuum........................750
Clock radio 5
Clothes washer.......................1450
Dryer (gas)................................300
Electric blanket 200
Electric clock...............................4
Furnace fan 500
Garage door opener.................350
Heater (portable)...................1500
Iron (electric)..........................1500
Radio/phone transmit 40-150
Sewing machine.......................100
Table fan 10-25
Waterpik 100

Refrigeration
Refrigerator/freezer 540
  22 ft³ (14 hrs/day)
Refrigerator/freezer 475
  16 ft³ (13 hrs/day)
Sun Frost refrigerator...............112
  16 ft³ (7 hrs/day)
Vestfrost refrigerator/ 60
  freezer 10.5 ft³
Standard freezer.......................440
  14 ft³ (15 hrs/day)
Sun Frost freezer......................112
  19 ft³ (10 hrs/day)

Kitchen appliances
Blender 350
Can opener (electric) 100
Coffee grinder 100
Coffee pot (electric)...............1200
Dishwasher.............................1500
Exhaust fans (3)........................144
Food dehydrator.......................600
Food processor.........................400
Microwave (.5 ft³) 750
Microwave (.8 to 1.5 ft³)........1400
Mixer 120
Popcorn popper 250
Range (large burner).............2100
Range (small burner).............1250
Trash compactor 1500
Waffle iron.............................1200

Lighting
Incandescent (100 watt)...........100
Incandescent light 60
  (60 watt)
Compact fluorescent 16
  (60 watt equivalent)
Incandescent (40 watt) 40
Compact fluorescent 11
  (40 watt equivalent)

Water Pumping
AC Jet pump (¼ hp).................500
  165 gal per day, 20 ft. well
DC pump for house 60
  pressure system (1-2 hrs/day)

DC submersible pump 50
  (6 hours/day)

Entertainment
CB radio 10
CD player....................................35
Cellular telephone 24
Computer printer......................100
Computer (desktop) 80-150
Computer (laptop)................20-50
Electric player piano 30
Radio telephone 10
Satellite system (12 ft dish) 45
Stereo (avg. volume) 15
TV (12-inch black & white) 15
TV (19-inch color)......................60
TV (25-inch color)....................130
VCR 40

Tools
Band saw (14")......................1100
Chain saw (12")......................1100
Circular saw (7¼") 900
Disc sander (9")......................1200
Drill (¼")................................250
Drill (½")................................750
Drill (1")................................1000
Electric mower 1500
Hedge trimmer 450
Weed eater 500

appliances, stereos, tools, etc. — you will be using in your new home. Once you've determined the amount of electricity used in a week, divide by seven to determine the average daily load in watt-hours.

Fig. 4.1: *The cleverly named Kill A Watt meter (a) and Watts Up? meter (b) are valuable tools for measuring the wattage of electronic devices and also ferreting phantom loads.*

DAN CHIRAS

a

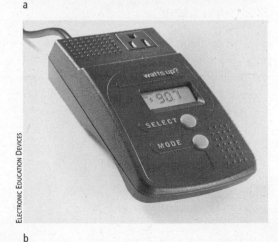

ELECTRONIC EDUCATION DEVICES

b

Although wind and solar assessors commonly use worksheets to calculate energy consumption for new homes and offices, the process leaves much to be desired. One shortcoming is that most of us haven't the foggiest idea how many hours or minutes the electrical devices in our arsenal of electronic gadgetry operate on a typical day. For example, how many hours do you operate your toaster each day? Is it three minutes, five minutes, or ten? And how often and how long do you run the blender? Is it ten minutes a day or fifteen? How long is the kitchen light on? How long does the refrigerator run? And what about the kids? Are they leaving lights on when you're not around? (Doubtful, but it is a possibility.) How many hours of television do they watch? How many hours a day do they spend on their computers?

Another problem is that electric consumption varies by season. Electrical lighting, for example, is used much less in the summer than the winter, because days are longer in the summer and people often spend more time outdoors when the weather's nice. In addition, furnace blowers operate a lot during the winter, but very little in the late spring and early fall and not at all in the summer. So what's the average daily run time of the furnace blower?

Another problem with this approach is that many electronic devices draw power when they're off. These are known as phantom loads. Television sets, VCRs, satellite receivers, cell phone and laptop computer chargers, automatic garage door openers, and a host of other devices all draw power after you've turned them off — sometimes nearly as much as when they're operating. Phantom loads, discussed in more detail in Chapter 8, typically account for five to ten percent of the monthly electrical consumption in US homes and, therefore, can add to the overall load, although they are rarely factored in to a load analysis.

Because of these problems, homeowners often grossly underestimate their consumption of electricity. If you use this technique, be careful.

To avoid underestimating electrical demand, it's a good idea to compare projected energy use to energy consumption in your existing home or business. If your estimate and actual energy usage differ considerably, and you can't explain why, you may want to start over.

Watt-hour meters, described earlier, can also be used. They keep track of how long devices run and will give you a more precise measurement of watt-hours. To accurately assess electrical consumption, plug devices like the refrigerator into the meter and leave them plugged in for a couple of weeks. Scott Russell argues, "A Watt-hour meter is the only way to obtain accurate consumption information for such loads."

Watt-hour meters range in price. The Kill A Watt meter costs about $30. The more expensive Watts Up? costs more than $100.

Once you've projected electrical energy use, it's time to determine whether a wind machine at your site will meet your needs, right?

Actually, no.

Before you proceed, we highly recommend that you consider energy-efficiency measures — that is, finding ways to slash electrical demand.

Energy efficiency measures help reduce load. The lower your energy demand, the smaller your wind-electric system will need to be. The smaller the system, the less money you'll have to put out initially.

As a general rule of thumb, Richard Perez, founder of *Home Power* magazine, asserts that every dollar invested in energy efficiency saves three to five dollars in system cost. In some cases, it can save even more. Therefore, if you invest $1,000 in measures that trim the electrical energy use in your home or office, you could save $3,000 to $5,000 in the cost of a renewable energy system. Robert Aram likes to tell his clients, "It's easier and cheaper to save a watt than to make a watt." Ian adds, "Money spent on energy efficiency measures, such as installing more efficient appliances, has a much better return on investment than money spent on a larger generating system." In addition, "Eliminating energy waste is also a smart move for the environment. While the fuel for renewable energy systems is available daily at no charge and no environmental impact, the collection equipment requires natural resources and the manufacturing processes do have an environmental impact. The smaller your system is, the less impact it will have."

So, before you determine which wind generator will meet your needs, we recommend that you develop an energy conservation and efficiency strategy. The best way is through simple behavioral changes such as turning off lights and turning down thermostats in the winter. We refer to these as energy conservation measures. ("These may be the most difficult to enact, as they involve changing personal habits," notes Mick.) The next most cost effective is developing ways to use electricity more efficiently. We'll take up both topics in the next section.

Conservation and Efficiency First!

Want to save 50 percent on the cost of your renewable energy system? Virtually all conscientious renewable energy experts agree: before a homeowner invests a single penny in a renewable energy system such as wind or solar electricity, the first step should be to make their homes — and their families — as energy efficient as possible. Even if you're frugal, you could be using a lot more electricity than you think and a lot more than you need to achieve the comfort and convenience you desire. Before you determine if the wind — or wind and some other renewable energy resource — can meet your electrical needs, take time to trim the fat out of your electrical energy diet.

Waste can be slashed in many ways, as most readers know. Interestingly, however, the ideas that first come to mind for most homeowners tend to be the most costly: energy-efficient washing machines, dishwashers, furnaces, air conditioners, and our perennial favorite, new windows. While all of these ideas are important, perhaps essential, to creating a more energy-efficient way of life or business, they're the highest fruit on the energy-efficiency tree — and the most expensive.

Before you spend a ton of money on new appliances, we recommend that you start with the low-hanging fruit. They are the simplest and cheapest modifications in a business or residence that yield the greatest energy savings per dollar invested. They yield the most gain with the least amount of pain.

Insulation, weatherstripping and caulk are three ideas for starters. All three make it easier to heat and cool homes and businesses. Caulking tiny leaks in the building envelope, applying weatherstripping around doors and windows,

> Before investing a single penny in a renewable energy system, such as wind or solar electricity, make your family and your homes as energy- efficient as possible.

Fig. 4.2: *Energy-efficient appliances like this Frigidaire Gallery horizontal axis (front-loading) washing machine decrease the consumption of electricity and offer the best return on investment. When contemplating a renewable energy system, be sure to make your home as efficient as possible first.*

FRIGIDAIRE

and adding insulation to walls and ceilings — even under floors over unconditioned spaces — reduce heat loss in the winter, keeping you and your family warm. They also reduce heat gain in the summer, resulting in cooler interiors. The increase in comfort is accompanied by a decrease in fuel bills — saving families and businesses heaps of money.

These measures not only save money on heating oil and gas — they also cut electrical use in homes heated by fossil fuels by reducing the run time of blowers on furnaces and pumps on boilers. In the summer, these measures reduce the run time of ceiling fans, air conditioners and evaporative coolers, saving even more energy.

All of these measures fall within the realm of energy conservation. Important as they are, there are even lower-hanging fruit. Huge savings can be achieved by simple changes in behavior. You've heard the list a million times. Turning lights, stereos, computers and TVs off when not in use. Turning the heat down a few degrees and wearing sweaters and warm socks in the winter. Turning the thermostat up and running ceiling fans in the summer helps save lots of energy, too. Opening windows to cool a home naturally, especially at night in regions that enjoy cool evenings, then closing the windows in the morning to keep heat out during the day. Drawing the shades or blinds on the east, south, then west side of your house as the summer sun moves across the sky. These changes cost nothing, except a little of your time, but reap enormous savings.

While many simple and easy measures can help you cut your electric consumption, addressing some big-ticket items can also make a huge dent in your monthly energy use. One of the biggest big-ticket items is the refrigerator. In many households, the fridge consumes the greatest amount of electricity. In fact, refrigerators may be responsible for a staggering 25 percent of electrical consumption in a home not heated or cooled electrically. So, if your refrigerator is old and in need of replacement, recycle it, and buy a new energy-efficient model. Thanks to dramatic improvements in design, refrigerators on the market today use significantly less energy than refrigerators manufactured 10 to 15 years ago — and they are quite affordable. Whatever you do, don't lug the old fridge out to the garage or take it down to the basement and use it to store an occasional case of beer. It will rob you blind!

Waste can also be reduced by installing energy-efficient electronics, such as Energy Star televisions and stereo equipment (Figure 4.3). And, of course, you can trim more fat from your energy bill by installing energy-efficient lighting, such as compact fluorescent lights, a technology that is now firmly entrenched in the market and soon to be the mainstay of lighting in the United States and other countries, such as Australia.

Although efficiency has been the mantra of the energy advocates for many years, don't discount its importance just because the advice is a bit threadbare. Few people, it turns out, have heeded the calls for energy efficiency. And many who have may not have fully tapped the potential savings. One of Dan's clients recently discovered that an insulation company that

Fig. 4.3: *Energy Efficient Stereo with Energy Star Label. This label assures you that the product you're looking at is one of the most energy efficient in its category.*

For an up-to-date list of energy-efficient appliances, US readers can log onto the EPA and DOE's Energy Star website at energystar.gov. Click on appliances. Consumer Reports has an excellent website that lists energy-efficient appliances as well. Their site also rates appliances on reliability, another key factor. Canadian readers can log on to oee.nrcan.gc.ca/energystar/english/consumers/ index.cfm for a list of international Energy Star appliances.

Assessing Your Wind Resource

Once you have executed a strategy to make your home or business — and its occupants — more energy efficient and have recalculated energy consumption, it is time to assess the potential of your wind resource. How much wind is available to you and when it is available? From this information, you can select a wind turbine that will produce enough electricity to meet your needs. After that, you can evaluate system costs and run a financial analysis to determine the return on investment.

retrofitted his home a few years ago only put in half the attic insulation that is needed today to slash his energy consumption. Because of these and other factors, there's a lot that homeowners and businesses can do to cut demand — often substantially. Huge economic savings are available for very little effort to those who pay heed to this message.

Those interested in learning more about making their homes energy efficient may want to read the chapters on energy conservation in one of Dan's books, for example, *Green Home Improvement*, which describes numerous projects to reduce energy use in homes and businesses. Another valuable book is the *Consumer Guide to Home Energy Savings* published by the American Council for an Energy-Efficient Economy and *Home Energy* magazine. This book contains lots of sound advice.

An assessment of available wind resources can be made using one or a combination of the following techniques: (1) direct measurement, (2) local airport and weather service data, (3) wind maps, and (4) online sources. You can also use other criteria to support your assessment, for example, the deformation of trees. This can help you confirm data from other sources and will be discussed shortly. Although all of these tools are important, professional small wind site assessors rely principally on state wind maps and online sources. They may use other measures to confirm their estimates.

Direct Measurement

By far the most accurate way of assessing wind is direct monitoring — measuring the wind speed at the tower height on the site you are considering. "Direct monitoring by a wind resource measurement system provides the clearest picture of the available resource," note the authors of the US Department of Energy's booklet, *Small Wind Electric Systems*. For best results, they contend, wind speed should be monitored over a period of at least one year. Given the variability of wind from year to year, however, the most useful results are obtained by two or more years of monitoring.

The best way to measure wind resources on-site is by using a pole-mounted anemometer located on the site at the height of the future wind turbine. Anemometers come in two basic varieties. A cup anemometer consists of small cups mounted to a central hub (Figures 4.4a and 4.4b). The other version is a propeller anemometer. It consists of a miniature wind machine (Figure 4.4c).

Anemometers and towers can be purchased online or can be rented or borrowed from government agencies in some states. Ideally, anemometers should be mounted at the same height as the future wind turbines. As a rule, an anemometer, like a wind machine, should be mounted so that it is at least 30 feet above the closest obstruction within 500 feet.

To assess wind speed at a site, you'll also need some way to record data, for example, a data logger. APRS World in Winona, Minnesota, sells an anemometer and a data logger. The data logger records speed and direction as well as time and date. Their data logger stores data on a removable 4-gigabyte card that can store a month's worth of data. The card is removed periodically and plugged into a card reader that plugs into a USB port on a computer. The data can then be uploaded into spreadsheets like Microsoft Excel and graphed and analyzed. The company also provides web-based software for analysis. You upload the raw data from the card on to their software that plots the data and provides basic statistics.

Other data loggers are also available. Another low-cost system is the Novalynx model 200-WS-25.

Although we don't recommend direct measurement for reasons that will be clear shortly, if you want to measure your wind resource and don't have the money, contact your state energy conservation office. They may have anemometers and towers that they lend to individuals interested in assessing the wind resource at their sites. Not all states offer this service, and, in those that do, availability is limited. You may find that the state only has four or five anemometers. If they're tied up for a year or two, you may be put on a waiting list. In addition, most states provide short towers, which will very likely provide you with useless data for sites with trees or other ground clutter.

Although direct measurement using a recording anemometer is considered the best way to determine average wind speed, these devices are not cheap. The anemometer typically costs $400 or more. You'll also need to add the cost of an appropriately sized tower and guy cables to secure the tower. With tower

Fig. 4.4: *Anemometers. (a) Cup anemometer. (b) Bob Aram installs a cup anemometer on a fixed lattice tower at a workshop sponsored by the Midwest Renewable Energy Association. (c) Propeller anemometers like this one are commonly used to measure wind speeds by the US Weather Service.*

costs, direct measurement could easily cost $5,000 or more. As a result, Mick notes, "No one in the small wind industry puts up anemometers anymore."

Airport or Weather Bureau Data

Another possible source of data on wind speed is a nearby airport or weather bureau. Airports and weather bureaus have been collecting data on wind speed regularly for ages. Unfortunately, this data has some serious shortcomings. One of them is that airport and weather bureau data is typically collected by anemometers on

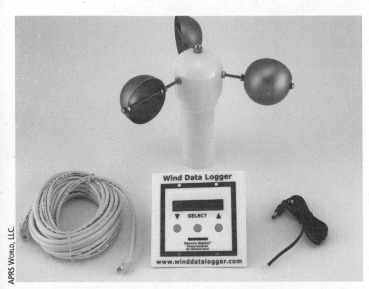

APRS WORLD, LLC.

a

DAN CHIRAS

b

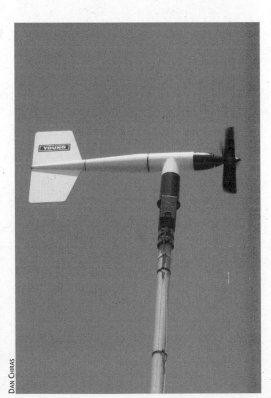

DAN CHIRAS

c

short towers. It is then adjusted to the 30-foot standard for reporting using some simple math, presented in the box on page 90.

While this approach sounds reasonable, it turns out to be a rather poor way of estimating wind speed. One reason is that many airport anemometers are installed in terrible locations. For instance, they may be mounted next to a hangar or on the roof of a building. They may be located next to trees or in depressions where they are sheltered from the winds. In some cases, they are mounted near roadways where they are affected by cars and trucks racing by. Ground drag and turbulence at these sites greatly diminish the wind speed. Extrapolation of these artificially low values from anemometers that are 10 feet off the ground to the values expected on a 30-foot tower, therefore, generally underestimate the wind resource. Although stronger winds might be present at the greater heights at which one would mount a wind machine, you simply can't tell from the data.

The second problem with airport wind speed data is that airports are frequently sited in the least-windy locations to ensure safer take-offs and landings. Small airports, for instance, are often sheltered by trees to reduce surface winds. Therefore, wind speed measurements recorded at airports could underrepresent the wind potential of a nearby site. The same is true for cities where most people live and where weather bureau data is collected.

The third problem with airports is that airport personnel are not trained to monitor wind, and many sites have wind monitoring equipment that is not maintained well or is not working properly.

The fourth — and most critical problem — comes from trying to extrapolate the 30-foot data to the hub height at which a wind turbine should fly. Extrapolating wind speed up from one data point is iffy at best. This technique is little better than "wet finger prospecting."

These shortcomings greatly reduce the value of airport data. In fact, as Mick puts it, airport data makes a lousy yardstick for determining wind speed at nearby sites. He adds, "Their average wind speeds are in all likelihood very low baseline numbers, really just a starting point for consideration. Virtually all wind generator sites I have seen have higher wind speeds by at least a mile per hour or two." If you don't think that two miles per hour is that much, remember that there is twice the available power in a ten mile per hour average wind speed than in an eight mile per hour average wind speed — that is, there's a 100 percent increase in available power. Many times the disparity is three or four miles per hour.

Although airport and weather bureau data is widely available, we believe it should be used as a last resort, and only if you live very near the airport and the topography and ground clutter of your site is very similar to the airport's.

Far superior data is available on state wind resource maps and on the Internet.

State Wind Maps

Because airport data and weather bureau data have their flaws and direct measurement is costly and time consuming, most professional wind site assessors get the data they need from other sources. One of the best and easiest ways to assess the wind resources in a region is to consult a state wind map.

State wind maps list average annual wind speed in meters per second and miles per hour. Unfortunately, wind speed estimates on the map are typically reported at 165 and 195 feet (50 or 60 meters) above the surface of the ground.[1] Most residential wind machines, however, are mounted at about 80 to 120 feet, sometimes as low as 60 feet (25 to 48 meters). Fortunately, a wind site assessor can use the data to extrapolate downward. The accompanying textbox explains how to do this.

Wind maps have come a long way in the past two decades and most states have good ones. You can locate your state's wind map at a couple of different websites. The Wind Powering America web site of the Department

Estimating Wind Speed from a State Wind Map

State wind maps provide annual average wind speed data but it is measured considerably higher (165 to 195 feet or 50 to 60 meters) than most residential wind towers (80 to 160 feet or 24 to 48 meters). Fortunately, average wind speed data from state wind maps can be adjusted using a relatively simple equation: $V = (H/H_O)^a V_O$.

In this equation, H is the height of the tower on which the turbine will be mounted. H_O is the height at which the average wind speed data is reported. V is the speed of the wind at the desired tower height, which is unknown. V_O is the average annual wind speed recorded on the state wind map. The Greek symbol α, or Alpha, is the wind shear coefficient. The shear coefficient is a measure of the change in wind speed with height above the ground caused by surface "texture" or "roughness," which is caused by ground clutter such as trees and buildings. It is used to make adjustments in wind speed at different heights based on the texture or roughness of the ground surface.

As noted in Chapter 2, the rate at which wind speeds increase with height varies based on the ground surface, notably the type of vegetation and the topography. The greatest increases occur over rough terrain, that is, terrain with numerous obstacles, such as trees and shrubs. The rate of increase is the lowest over smooth terrain, for example, a short-grass prairie or a meadow. Table 4.3 lists typical shear coefficients.

To estimate wind speed at a proposed tower height, simply fill in the numbers and do the math. You will need a calculator with the exponent function, which all but the simplest ones have. To see how this works, suppose that average wind speed from the state wind map at 198 feet at a rural site with scattered trees and building is 15 miles per hour. The wind shear coefficient is 0.24, according to Table 4.3.

To calculate the average wind speed at 120 feet, we begin with the equation:
$V = (H/H_O)^a V_O$ ☞

of Energy is at windpoweringamerica.gov/index.asp. It's probably faster to search for "Wind Powering America" than to type out this URL. This site contains regional and state maps with wind speeds at 165 to 195 feet (50 to 60 meters). Ian finds maps at www.awstruewind.com. AWS Truewind prepared most of the Wind Powering America maps. The *Canadian Wind Atlas* can be obtained by logging on to cmc.ec.gc.ca/.

While wind maps are an excellent source of information for your area, they do have some limitations. One of them is resolution. In some areas, like the plains of western Kansas, wind maps show uniform wind speeds over large areas. If you live in one of those areas, the map will give you a pretty accurate idea of average wind speed. In other areas, however, the topography is much more complex. Wind speeds may vary considerably over short distances in

To solve the equation, plug in the values.

$$V = (120/198)^{.24} \times 15 = (0.61)^{.24} \times 15 = 13.3 \text{ mph}$$

The average wind speed at the proposed height is about 13.3 miles per hour. Professional wind site assessors round down to be conservative. They'd use 13 miles per hour to determine the electrical production of a wind turbine.

When using this equation, keep in mind that the height measurements must be in the same units.[2] Both must be in feet or both must be in meters. Similarly, the wind speed must be in consistent units, either miles per hour or meters per second. Don't mix units or the results will be wrong.

To calculate exponents, you can use a calculator. Google also provides a very convenient calculator. Simply type in the equation in the search window and hit the search button. ■

Table 4.3 Typical Wind Shear Coefficients	
Surface Characteristics	**Shear Coefficient**
Lake or ocean, water or ice	0.10
Short grass or tilled ground	0.14
Level country, foot high grass, occasional tree	0.16
Tall row crops, hedges, short fence rows	0.20
Hilly country with open ground	0.20
Few trees, occasional buildings	0.22
Many scattered trees, more buildings	0.24
Wooded country, small town	0.28
Suburbs	0.30
Urban areas	0.40

such terrain and the resolution of the maps isn't good enough to pinpoint an exact location. Dan's home in the foothills of the Rocky Mountains is a good example. The state wind map shows three different classes of winds in his area. Which one you're in depends on where you live. A neighbor half a mile away in a valley may have very little wind, while a neighbor perched on top of one of the many mountains nearby will have a great deal of wind. While this seems like common sense to someone experienced in siting wind turbines, it often goes unnoticed by newcomers to wind.

In assessing wind resources, location is everything. Topography can confuse matters considerably. So, just because a map shows that wind speeds in your area are sufficient, it doesn't mean they are. Hills, cliffs, forests and buildings can deflect winds, greatly reducing wind flow at downwind sites. Some types of hills and cliffs can even magnify winds. So, the more complex the terrain, the less accurate the wind maps are.

Wind maps are adequate in most cases for siting a small wind turbine. That said, it's always a good idea to study the topography, vegetative cover, and ground clutter when estimating average annual wind speed.

Using data from wind maps, it is possible to make an initial, approximate determination regarding the suitability of a site. But how much wind do you need to make a system worth the investment?

In Mike Bergey's online article, "A Primer on Small Turbines," he states that a stand-alone (off-grid) wind energy system makes sense if you're in an area with an annual average wind speed of eight miles per hour or higher at hub height. It should, he contends, be economically feasible, to which Mick adds: especially for those whose off-grid systems incorporate PV. That said, Ian points out that many people choose to make wind electricity with lower wind speeds. He and his family have lived for more than 20 years in an area with 7+ mile-per-hour annual average wind speeds and wind turbines have supplied about half of his family's year-round electricity and a high percentage of his family's winter energy. Without the wind generators, the Woofendens would have been using a lot of fossil fuel over the years.

While an 8-mile-per-hour average wind speed may suffice for an off-grid system, according to Bergey, grid-connected systems typically require a higher average annual wind speed of about 10 miles per hour to be economical. Again, this speed is at the hub height of the turbine.

Dan advises caution when using any general recommendations. With the high cost of wind turbines nowadays, he's found that higher wind speeds — 12 miles per hour or higher — are necessary to make a wind system economical.

Online Databanks

While state wind maps are an excellent source of information, not all states have them or have invested in the best mapping technology. Because of this, some professional small wind site assessors perform small wind site analyses using a very extensive online database developed by NASA. This website, Surface Meteorology and Solar Energy, is at eosweb.larc.nasa.gov/sse.

Surface Meteorology and Solar Energy provides a wealth of data on wind energy. It contains tables that show both the monthly and annual average wind direction and average wind speed at sites anywhere in the world.

Although the site is very user friendly, a few tips on navigating your way through to the data may save you some brain damage. The accompanying sidebar provides detailed instructions on how to use the site, starting with the account setup. By the way, there's no fee to set up an account and it only takes a few seconds.

Like wind maps, NASA data provides good information, but you may still need to assess your site very carefully, especially in complex topography. If your site is in a valley or hilly terrain, for instance, the average wind speed data may not represent the average wind speed at your site. If the proposed turbine is downwind from numerous buildings, the wind speed at the wind generator may be lower that the satellite data suggests.

When in doubt, hire a professional wind site assessor to analyze your site and make recommendations for tower/turbine placement and minimum acceptable tower height. You can find a list of certified wind site assessors online at the Midwest Renewable Energy Association's website, the-mrea.org. They're the national certifying body for wind site assessors. You may also want to ask a professional installer to render an opinion, although not all installers are as knowledgeable about wind and wind site assessment as certified wind site assessors. Some installers may be more interested in selling and installing a turbine and tower than in giving you an objective wind site assessment.

Flagging and Other Factors

Even if a state wind map or the NASA data suggests the presence of a good resource on a site, you need to study other factors such as the topography, vegetative cover, the proximity of the wind turbine to trees, hills, cliffs and buildings, all of which could decrease — and, in some cases, increase — wind speeds. You also have to take into account local wind patterns like mountain-valley winds and offshore and onshore winds.

In some areas, vegetation provides an important clue as to how windy a site is. Figure 4.6 shows a tree on a client of Dan's property in the wind-swept foothills of the Rockies. This

Fig. 4.5: *Effect of Topography on Wind Speed. Hills can dramatically increase wind speed. Placement of a tower on the top of the hill could result in a significant increase in the power output of a turbine. Placement of a turbine at the base of a hill could result in much lower output.*

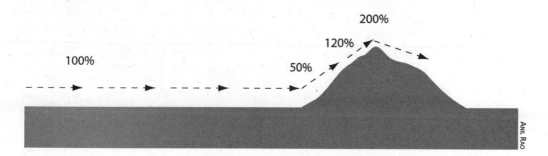

100%   50%   120%   200%

Anil Rao

How to Tap into the NASA Surface Meteorology and Solar Energy Database

1. Log on to eosweb.larc.nasa.gov/sse.
2. After logging on to the site, click on "Meteorology and Solar Energy."
3. Then either click on "Enter the Latitude and Longitude" or the map of the world to locate your site (the latter is slower).
4. A new screen will appear. It will ask you to enter your e-mail address and password. If you haven't set up an account, do it now.
5. After setting up an account, enter the latitude and longitude of your site.
6. The latitude and longitude of a site can be determined by calling a local survey, checking the gazetteer for your state, referring to a topographic map, or checking out a satellite image of a property online. Aerials that include latitude and longitude are available online at terraserver-usa.com.
7. Entering latitude and longitude is one place where confusion may occur. Latitudes and longitudes may be entered as decimal degrees (33.5) or as degrees and minutes, separated by a space (33 30). Latitudes are either north or south, longitudes are either east or west. South latitudes and west longitudes are both assigned a minus sign to avoid confusion.
8. Once you have entered the latitude and longitude, click on "Submit."
9. You'll now be presented with a long list of data from which to choose. You can confirm that you've entered the right latitude and longitude by clicking on the words, "Show a location map." When you do, it will present a map showing the location you've selected. If it's not correct, check the latitude and longitude or check to be sure you entered it correctly.
10. Scroll down to the first box on wind, entitled "Meteorology (Wind)." In this box, you can click on one of several options. I usually click on "Wind direction at 50 meters" to determine the predominant wind flow at a site. This number will be expressed in degrees, and is the direction the wind is coming from. Zero degrees is north. Ninety degrees indicates wind from the east; 180 degrees indicates wind from the south, while 270 degrees indicates wind from the west.
11. In the box below, click on "Wind Speed at Any Height and Vegetation Type."
12. Now scroll through the list to select a vegetation type that represents the site you want to assess. This will enter a shear factor or alpha, discussed earlier in this chapter. Choose the selection that best describes your site.
13. Now enter height of the proposed wind tower in meters. To convert feet to meters, divide by 3.3.
14. Now scroll to the bottom of the page and click on "Submit."
15. You'll be presented with the data you requested, in this case the predominant wind direction and average wind speed data both by month and by year. Note that the data is presented in meters per second. To convert meters per second to miles per hour, multiply by 2.237.
16. Data tables can be cut and pasted into documents for reports or for your own use later on.

Fig. 4.6: *Flagging. The pine tree in the distance on a windy site in the foothills of the Rockies in Colorado exhibits moderate flagging, indicating an average annual wind speed of 11 to 13 miles per hour. The pine tree in the middle exhibits complete flagging, indicating 13 to 16 mile-per-hour average annual wind speed. Both trees are on the same property, illustrating how wind can vary within a short distance.*

tree exhibits a phenomenon known as flagging. What is flagging?

As readers know, a flag blowing in the wind blows to one side of the pole. Strong winds produce a similar but permanent visual effect in trees, especially coniferous trees, called flagging. When flagging occurs, the branches of a tree appear to stream downwind. Persistent strong winds do not cause the branches of a tree to bend around and fly on one side of the trunk. Rather, they damage or stunt the upwind branches, so it appears as if the downwind branches stream downwind. In really strong winds, the upward branches may be stripped away entirely.

Flagging can be used to approximate the average wind speed at some sites, as shown in Figure 4.7. This scale is known as the Griggs-Putnam Index. According to this index, slight deformity, known as "brushing and slight flagging," indicates a probable mean annual wind speed in the range of seven to nine miles per hour. Slightly more flagging, cleverly referred to as "slight flagging," indicates an approximate average annual wind speed of 9 to 11 miles per hour. Moderate flagging indicates an average annual wind speed of 11 to 13 miles per hour, and so on. On the extreme upper end, the trees are bent to the ground, a phenomenon called carpeting.

Fig. 4.7: *Griggs-Putnam Index of Deformity*

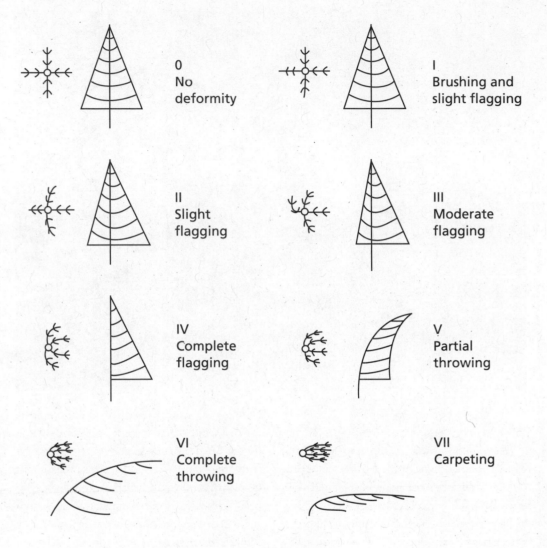

0
No deformity

I
Brushing and slight flagging

II
Slight flagging

III
Moderate flagging

IV
Complete flagging

V
Partial throwing

VI
Complete throwing

VII
Carpeting

Griggs-Putnam Index of Deformity

Index	I	II	III	IV	V	VI	VII
Wind Speed (mph)	7-9	9-11	11-13	13-16	15-18	16-21	22+

Flagging is typically observed in single trees — that is, a tree growing alone in a field isolated from other trees and obstacles. Trees in forests or groves are not prone to flagging except at their tops.

Flagging can be used to verify data from NASA or wind maps. If the wind map or NASA data indicates that a site has an average wind speed of 15 miles per hour and the trees exhibit complete flagging, you can feel confident that the data is correct.

Although flagging helps verify other data, it is important to note that the absence of flagging does not necessarily indicate a lack of wind. Some trees aren't very susceptible to flagging. Most deciduous trees, for example, are less prone to flagging than coniferous trees.

Site assessors and homeowners can also look for other telltale signs of strong winds on a site. If neighbors or local governments use snow fences to prevent snow from drifting over driveways and roads, it's a good indication of strong winter winds. Clouds of dust and dirty snow are signs of soil erosion by wind, a good sign for wind site assessors (a bad sign for farmers). The use of windbreaks, trees planted upwind from homes and farmsteads to protect them from winds, is another good sign that there's wind to be reckoned with. Tattered flags flying on flagpoles tell the same story. Telephone and electric poles that tilt at an angle convey a similar message.

How Much Electricity Will A Wind Turbine Produce?

Once you've determined the average wind speed at a site, it is time to determine how much electricity a wind generator could produce at the proposed tower height — and therefore whether it can meet all your needs or what percentage of your needs it will satisfy. This step is fairly easy.

Table 4.4 shows a list of wind turbines and the estimated annual output of each turbine (in kilowatt-hours) at seven different average wind speeds. (The wind turbines listed in the table are some of those for which Wisconsin's Focus on Energy program offers rebates.) To see how this table is used, consider an example.

Let's assume that the average wind speed at a site (at hub height) is 12 miles per hour. Let's also assume that your load analysis, after efficiency measures have been implemented, indicates you'll need, on average, 900 kilowatt-hours per month, or 10,800 per year. In the 12 mile-per-hour column, you'll discover two wind turbines that match your electrical requirements, Proven WT6000 and Bergey Wind Power's XL-S. If the wind speed at your site is 14 miles per hour, an Endurance wind turbine would meet your needs. The Proven WT 6000 and the Bergey XL-S would produce more than you need.

Annual energy outputs used to estimate the economic performance of a wind energy system can also be found in an article Ian and Mick published in *Home Power* magazine (Issue 122), entitled "How to Buy a Wind-Electric System."

You can also obtain annual energy output data directly from wind turbine manufacturers. While this data is useful, we think that manufacturers tend to overstate the electric

Table 4.4 Kilowatt-Hour/Year Outputs							
Wind in miles per hour	**Swept area**	**8**	**9**	**10**	**11**	**12**	**13**
Whisper 100	39	360	540	780	960	1,200	1,500
xl 1	53	660	1,020	1,380	1,800	2,256	2,640
WT 600	55	504	792	996	1,356	1,488	1,752
Whisper 200	63.5	720	1,080	1,500	1,920	2,280	2,700
WT 2500	97	2,004	2,472	3,516	3,996	5,004	5,580
ARE 110	110	1,620	2,316	3,144	4,068	5,040	6,060
Skystream	113	1,200	2,040	2,880	3,720	4,560	5,400
Whisper 500	176	2,040	2,760	3,960	4,920	6,456	7,440
Endurance	254	1,843	3,091	4,587	6,268	8,068	9,920
WT 6000	254	5,004	6,768	8,004	11,004	12,996	15,000
BWC XL-S	415	2,880	4,440	6,240	8,400	10,800	13,560
ARE 442	442	7,476	10,440	14,052	17,640	21,972	25,584
Jacobs 31-20	754	9,828	13,920	19,728	25,704	32,292	39,288
V-15-35	1,964	N/A	38,000	43,000	58,000	64,000	80,000
PGE 20/32	3,120	N/A	53,280	64,920	82,296	90,000	107,796

MICK SAGRILLO

production of their turbines. As a result, we recommend derating their estimated outputs by 20 percent — just to be conservative. (The data from Focus on Energy is derated annual energy output; the data in Ian and Mick's article is manufacturer data and is not derated.)

Once you've found the wind turbines that meet your need at the proposed tower height, it's time to consider the cost.

Does a Wind System Make Economic Sense?

At least three options are available when it comes to analyzing the economic cost and benefits of a wind turbine: (1) a comparison of the cost of electricity from the wind turbine with conventional power or some other renewable energy technology, (2) an estimate of return on investment, and (3) a more sophisticated economic analysis tool known as discounting. We'll present an overview of each method in this chapter.

Cost of Electricity Comparison

One of the simplest ways of analyzing the economic performance of a wind energy system is to compare the cost of electricity from the wind system to the cost of electricity from a conventional source — notably, the local utility — or some other renewable energy technology you are considering. This is a five-step process, three of which we've already discussed.

First, determine the average monthly electrical consumption of your home or business. Second, determine the average monthly wind speeds at the site. Third, identify a wind turbine that produces a sufficient amount of electricity to meet your needs. If you are contemplating an off-grid system, you want a month-by-month match. If you're installing a utility-tied system, you might only want an annual match.

After estimating the monthly electrical production, convert this to annual electrical output. In the example presented earlier, a Proven WT6000 would produce 1,083 kilowatt-hours per month or 12,996 kilowatt-hours per year. Now multiply the kilowatt-hours produced in a year by the life of the system. How long would that be?

A well-made, heavy-duty wind turbine like the Proven WT 6000 could last 20 to 30 years, with regular inspection once or twice a year and maintenance and repairs as required. A lighter-weight and cheaper model might only last five years or fewer. If the Proven WT6000 lasts 30 years, it would produce about 389,880 kilowatt-hours over its lifetime. In 2007, the Proven on a 120-foot guyed tower cost about $60,000, installed, in the United States. Dividing the cost of the system by the total output yields the cost of electricity per kilowatt-hour. In this example, then, the electricity, over the lifetime of the turbine, will cost about 15 cents per kilowatt-hour. Is this a bargain?

It depends.

If the alternative source costs 15 cents per kilowatt-hour or more, the Proven turns out to be a pretty good investment. Although the turbine will require maintenance and repair over the years, the cost of electricity from the utility is also bound to increase. It's been rising, nationwide, at a rate of 4.5 percent per year for the 20 years prior to 2006 and could, many think, increase more rapidly as energy prices rise. As a result, these two factors could offset each other. (We'll cover maintenance and maintenance costs in Chapter 9.)

This system would make economic sense if you were paying 15 cents per kilowatt-hour or more for locally generated electricity, but only if you don't mind prepaying your electrical bill — laying down $60,000 all at once.

And, lest we forget, you need to have access to $60,000.

Or you'll need to be able to borrow it.

To make a fair analysis, you should factor in the cost of lost economic opportunities, the money you could have made on that $60,000 had you invested it in something else. Would the rise in electrical costs offset lost income? If you take out a loan, you have to factor in the cost of interest on the loan. You also have to factor in the cost of insurance (property and liability insurance, discussed in Chapter 10) and increased property taxes, if any, resulting from the installation of the turbine. (Some states waive property taxes on renewable energy systems.)

When calculating the cost of electricity from a wind system, don't forget to subtract financial incentives from state and local government or local utilities from the initial cost.

These incentives can be quite substantial. In Wisconsin, for example, over 30 utilities participate in a statewide program called Focus on Energy. They provide customers who install wind energy systems up to 25 percent of their system cost with a maximum reward of $35,000. New York State offers generous incentives as well, as do a few others, including New Jersey, Massachusetts, California and Oregon. Starting in 2009, the federal government will offer a 30% incentive to homeowners. The US Department of Agriculture offers a 25 percent grant to cover the cost of wind systems on farms and rural businesses. To learn more about incentives in your state, log on to the Database of State Incentives for Renewables and Efficiency, dsireusa.org.

How important are financial incentives such as these?

In many cases, wind system installations are driven by rebates or buydowns. In the example we've been using, the Proven WT 6000, a $60,000 turbine and tower would cost $45,000 with a 25 percent rebate. The cost of electricity over the life of the turbine would now drop to 11.5 cents per kilowatt-hour.

Last but not least, when building a new home, don't forget to include the cost of connecting to the electrical grid when comparing the cost of wind-generated electricity to the cost of power from the utility. As noted in the last chapter, if your home or business is more than a few tenths of a mile from existing electric lines, you could face a huge bill to connect to the grid. If your home is located even farther away, the cost could be higher. A half-mile line

could cost as much as $50,000. What is more, the cost of connecting to the electric grid only covers the cost of the installation of poles and electric lines. It does not include a single kilowatt-hour of electricity. As a result, if you're a building a new home that is more than a few tenths of a mile from a power line and the utility company requires customers to pay for line extension, an off-grid wind system that costs $60,000 may be well worth the investment. (Don't forget to add the cost of other equipment such as an inverter, gen-set, battery bank and periodic replacement of batteries in an off-grid system.)

Comparing the cost of electricity from a wind energy system to the cost from conventional power (or another source of renewable energy) is a very simple way to assess the economic feasibility. For the sake of simplicity, however, it does overlook a few factors, such as the rising cost of electricity, opportunity costs, and the cost of interest on loans.

To perform a detailed financial analysis, it is best to include *all* of the costs that you can reasonably predict and estimate. If you need help, you may want to hire an accountant to run the numbers for you. Or you can use the spreadsheet, which we'll introduce shortly. It will handle all of this for you, and allow you to compare more accurately the cost of a wind machine over the lifetime of the system to the cost of utility power.

Calculating Return on Investment

Another relatively simple way of determining the cost effectiveness of a renewable energy system is to determine the return on investment (ROI). Return on investment is, as its name implies, the rate of return expressed as a percentage of an investment.

ROI can be calculated by dividing the annual value of electricity generated by a wind system by the cost of the system. As a simple example, suppose a wind energy system produces $1,200 worth of electricity each year and costs $12,000 to install. The simple return on investment is $1,200 divided by $12,000, or 10 percent.

As another example, let's calculate the return on investment of the $60,000 Proven WT6000 used in our earlier example. If the turbine costs $60,000 installed and produces 12,996 kilowatt-hours per year and electricity from the utility costs 20 cents per kilowatt-hour, the electricity would be worth $2,599.20 per year. To calculate the return on investment, divide the annual value of electricity by the cost of the system ($2,599.20 divided by $60,000). In this instance, the return on investment would be 4.3 percent. Not terribly good, but not bad either, especially if you factor in all of the feel-good variables, like producing your own power from a clean, renewable energy source.

If this system were installed with a 25 percent buydown or rebate, the simple annual return on investment would be 5.8 percent, which is much better than a certificate of deposit at this writing (October 2008).

Like the previous method, simple return on investment does not take into account a number of economic factors that could influence one's decision. For instance, it does not include interest payments on loans required to purchase the system or lost interest if the turbine and tower are paid in cash. It does not take into account insurance costs or property taxes. All of these factors decrease the value of the investment.

This calculation also does not take into account factors that increase the value of this investment, for example, the potential increase in the cost of electricity from the local utility. This would make wind generated electricity more valuable. Moreover, this calculation does not take into account the fact that money saved on the utility bill is tax-free income to you. Nor does it take into account possible income tax benefits for businesses. Businesses, for instance, can use an accelerated depreciation schedule for renewable energy systems such as wind and solar electricity. Businesses also qualify for higher rebates from the federal government. They receive 30 percent on systems with no cap, which is applied to the purchase of residential systems.

This analysis also does not account for the potential use of electricity as a transportation fuel, for example, in electric cars or plug-in hybrids.

While electricity is typically used to power lights, appliances and electronics in a home and business, it can also be used to power vehicles.

According to various sources, seven kilowatt-hours of electricity is equivalent to one gallon of gasoline. As a source of light, that 7 kilowatt-hours of electricity would be worth 70 cents if you were paying 10 cents per kilowatt-hour. As a source of propulsion in an electric car, the 7 kilowatt-hours of electricity would be equal to $4 if gas were selling at $4 per gallon. (The electricity would then be worth nearly 60 cents per kilowatt-hour.)

Simple return on investment is a great way of evaluating the economic performance of a renewable energy system. It's light years ahead of the black sheep of the economic tools, payback. For reasons that will be clear shortly, we and other professionals avoid payback discussions at all costs when talking to clients or in the classes and workshops we teach.

Why?

As most readers know, payback is a term that gained popularity in the 1970s. It is used for energy conservation measures and renewable energy systems. More recently, payback has been applied to energy-efficient hybrid vehicles. Payback is, quite simply, how long it takes a system or energy conservation measure in years to pay back its cost through the savings.

For a wind energy system, payback can be determined by dividing the cost of the system by the anticipated annual savings. If the $60,000 wind energy system we've been looking at produces 12,996 kilowatt-hours per year and grid power costs you 20 cents per kilowatt-hour, the annual savings of $2,599 yield a payback of 23 years ($60,000 divided by $2,599). In other words, it will take the savings from your system 23 years to pay off the cost, ignoring the maintenance and repair costs. From that point on, the system produces electricity free of charge. If that same system received a rebate of 25 percent, the payback would be 17 years. Both are pretty daunting.

While 23-year and 17-year payback seem ridiculously long, the return on investment, calculated earlier, was actually 4.3 percent in the first case and 5.7 percent in the second, both fairly decent returns on investment. Interestingly, while the 5.7 percent return on investment seems fairly decent, the 17-year payback, which represents the same return, seems prohibitively long.

Simple payback is pretty straightforward and fairly easy for most of us to understand, but it has very serious drawbacks. The most important is that it is seriously misleading, as the previous example illustrates. While you might love a 10 percent return on your investment, a 10-year payback seems awfully long when, in fact, they're exactly the same, assuming we ignore opportunity costs, inflation and other factors when calculating both payback and return on investment.

Simple payback and simple return on investment are closely related metrics. Mathematically speaking, return on investment is the reciprocal of payback. That is, ROI = 1/Payback. As just noted, a wind system with a 10-year payback has a 10 percent return on investment (ROI = $\frac{1}{10}$).

One final word on the subject: it's worth noting that we rarely use payback, or ROI for that matter, when considering purchases in our lives. Do we calculate the payback of our bass boat costing $25,000, or the new car?

Discounting and Net Present Value: Comparing Discounted Costs

The best way to determine whether an investment in wind energy makes economic sense is to compare the cost of the system to the cost of electricity it will displace. This analysis, unlike the others we have presented in this chapter, takes into account maintenance costs and the rising cost of grid power. It also takes into account the time value of money — the fact that a dollar today is worth less tomorrow, and even less a few years from now. These calculations incorporate a factor economists call a discount rate that allows them to factor in the declining value of money, as we'll explain shortly.

To make life easier, this economic analysis can be performed by using a spreadsheet like the one shown in Table 4.5 provided by renewable energy economist John Richter of the Sustainable Energy Education Institute in Michigan. This spreadsheet is available for your use on Dan's web site, courtesy of John. We recommend that you use it.

To understand how this technique is used, consider an example. Table 4.5 shows the economic analysis for a 10 kilowatt-hour wind turbine that costs $32,000 to install. Grid power costs eight cents per kilowatt-hour. John assumed a maintenance cost of $100 every other year. Although this is low, we'll use the number simply to demonstrate the process. The turbine was installed in a class-5 wind site and was capable of producing 25,1000 kilowatt-hours per year at this site.

As you can see by examining Table 4.5, the spreadsheet contains several columns. The first column on the left is years — how far out you want to run the calculation. This analysis was run for a period of 20 years, the lifetime of the system.

The next column is the discount factor. For simplicity, Richter recommends choosing the highest interest rate on any debt you have, including your mortgage. Or, if you have no debt, choose the highest investment interest rate you can get with a risk profile similar to the renewable energy system, which is usually very low. A ten-year government bond is a good basis and usually pays about five percent. John used 4.5 percent in his calculation.

The next column (under the category "Buy Utility Electricity") shows the cost of electricity from the local power company — that is, how much you would be paying each year for electricity purchased from the local utility — taking into account rising fuel costs (4.5 percent annual increase). The bottom figure in the column is the total cost of electricity — $62,994.

How do you know what number to use here?

If your wind machine is going to produce all of your electricity, add up your monthly bills and enter this number in year one. In this example, the total was just over $2,000. If the wind machine will produce 80 percent of your electrical needs, based on your analysis of wind resources and the output of the wind machine on the tower, enter this amount. If it will produce 50 percent of your electricity, use this number.

To customize this analysis, check with your utility. They should be able to provide data on the average increase in their electrical rates over the past 10 to 20 years. You may want to use a shorter time frame because rates are rising more rapidly now.

The next column under "Buy Utility Electricity" is the discounted cost of electricity from the utility — the cost of electricity adjusted for the decline in the value of money. In this example, you will shell out $63,000 over 20 years to the utility company. "Because money now is more precious than money later," says John Richter, "we cannot simply sum all the expenses of electricity, and ignore when they occur." Richter adds, "Expenses ten years from now are not the same as expenses today. We must 'discount' those future expenses to

Year	Discount Factor	Buy Utility Electricity		Proposed Wind System	
Rate	5.2%	Cost 4.5%	Discounted Cost	Cost	Discounted Cost
0	1.000	$0	$0	$32,000	$32,000
1	0.951	$2,008	$1,909	$0	$0
2	0.904	$2,098	$1,896	$100	$90
3	0.859	$2,193	$1,883	$0	$0
4	0.816	$2,291	$1,871	$100	$82
5	0.776	$2,395	$1,858	$0	$0
6	0.738	$2,502	$1,846	$100	$74
7	0.701	$2,615	$1,834	$0	$0
8	0.667	$2,733	$1,822	$100	$67
9	0.634	$2,856	$1,809	$0	$0
10	0.602	$2,984	$1,797	$100	$60
11	0.573	$3,118	$1,785	$0	$0
12	0.544	$3,259	$1,774	$100	$54
13	0.517	$3,405	$1,762	$0	$0
14	0.492	$3,559	$1,750	$100	$49
15	0.467	$3,719	$1,738	$0	$0
16	0.444	$3,886	$1,727	$100	$44
17	0.422	$4,061	$1,715	$0	$0
18	0.402	$4,244	$1,704	$100	$40
19	0.382	$4,435	$1,693	$0	$0
20	0.363	$4,634	$1,681	$100	$36
Total		$62,994	$35,855	$33,000	$32,597

Table 4.5
Cost Comparison Class 5 Site

JOHN RICHTER

reflect the time-value of money. This gives us the 'present-value' of those future expenses."

The discounted cost of electricity in column four is therefore the present value of the money spent on electricity over the next 20 years. In this example, the present value of the money spent on electricity from the utility is $35,855.

This leads us to the final steps. In the fifth column of the spreadsheet, John entered the cost of the wind energy system — $32,000. This figure should include all of the costs described earlier in the chapter minus financial incentives you will receive. Notice in the example we're using there's a $100 maintenance expense every other year. The actual outlay will be $33,000.

The next and final column of the spread sheet is the discounted cost of the wind machine. Like the discounted cost of electricity, it is the present value of your expenditure. That is, it takes into account loss in value due to inflation.

The final step is a simple comparison of costs. Compare the discounted cost (net present value) of electricity from a wind turbine to the discounted cost (net present value) of electricity from the utility. In this example, the net present value in this exceptional wind site is $32,597, more than $3,000 below the cost of electricity. When performing such analyses, if the wind system is cheaper, it makes economic sense. If it costs more, it doesn't. If the differential is small, you may want to go for it anyway.

Table 4.6, shows a similar analysis, but in a class-3 wind site. In this instance, you can see that the discounted cost of electricity from the grid ($25,999) is substantially less than the discounted cost of the system ($32,597).

After running the spreadsheet or calculating payback and return on investment, you will be equipped with the information needed to make a rational economic decision about wind power. Don't forget, however, that there are other mitigating factors. As Richter pointed out at a workshop at the 2006 Michigan Energy Fair sponsored by the Great Lakes Renewable Energy Association, "Even though a system may not make perfect sense from an economic standpoint, it is your money. You can spend it how you see fit." You may, for instance, purchase a system for peace of mind — knowing you will be free from utility power and not subject to rising fuel costs. Being independent of the power company is often a compelling reason for independent-minded folks. Creating an opportunity to sell power to the local utility may be worth the investment. Or, you may find the personal satisfaction of generating power from a clean, renewable resource sufficient. Just having a fancy new toy to play with may also be worth a slightly higher electrical cost.

Putting It All Together

Now that you have seen how one determines whether a wind energy system makes economic sense, it should be clear that blanket statements about "qualifying" wind speeds — wind speeds that make a wind energy system worthwhile — are only general guidelines. Economics is where the rubber meets the road in many cases. Comparing wind against the "competition," calculating the return on investment, or comparing strategies based on net present value, gives a potential wind buyer a much more realistic view of the economic feasibility of wind energy at a particular site.

As we have already pointed out, economics is not the only metric by which we judge our decisions. Energy independence, environmental values, reliability, the "cool" factor, and the fun value of a hobby for the renewable energy

		Table 4.6 Cost Comparison Class 3 Site			
Year	Discount Factor	Buy Utility Electricity		Proposed Wind System	
Rate	5.2%	Cost 4.5%	Discounted Cost	Cost	Discounted Cost
0	1.000	$0	$0	$32,000	$32,000
1	0.951	$1,456	$1,384	$0	$0
2	0.904	$1,522	$1,375	$100	$90
3	0.859	$1,590	$1,366	$0	$0
4	0.816	$1,662	$1,357	$100	$82
5	0.776	$1,736	$1,348	$0	$0
6	0.738	$1,814	$1,339	$100	$74
7	0.701	$1,896	$1,330	$0	$0
8	0.667	$1,981	$1,321	$100	$67
9	0.634	$2,071	$1,312	$0	$0
10	0.602	$2,164	$1,303	$100	$60
11	0.573	$2,261	$1,295	$0	$0
12	0.544	$2,363	$1,286	$100	$54
13	0.517	$2,469	$1,277	$0	$0
14	0.492	$2,580	$1,269	$100	$49
15	0.467	$2,696	$1,261	$0	$0
16	0.444	$2,818	$1,252	$100	$44
17	0.422	$2,945	$1,244	$0	$0
18	0.402	$3,077	$1,236	$100	$40
19	0.382	$3,216	$1,227	$0	$0
20	0.363	$3,360	$1,219	$100	$36
Total		$45,677	$25,999	$33,000	$32,597

JOHN RICHTER

gear heads, among other factors, also play prominently in our decisions to invest in renewable energy. They are all perfectly valid motivations for installing a renewable energy system, and should never be forgotten.

As Ian points out, "We have certain values by which we strive to live, and we pay good money for things we value." People often invest in renewable energy because they want to do the right thing.

If you are committed to a renewable energy future, return on investment may be irrelevant. Creating a sustainable future, helping reverse costly global climate change, protecting the world for your children and your grandchildren, setting a good example, leaving a legacy, helping the fledgling wind energy industry take off, and doing the right thing are all perfectly acceptable reasons for installing a wind system.

CHAPTER 5

A Primer on Wind Generators

With approximately 50 companies worldwide producing small wind generators, you'll discover a wide choice of models and many factors to consider when it comes to buying a wind turbine. This chapter will help you understand your options. We'll begin with the basics by examining the types of wind generators on the market today and then examine the components of the most common turbines, horizontal-axis machines. We'll also examine vertical-axis wind turbines and dispel the common myths about them. To help you sort through your options when shopping for a turbine, we'll discuss important features, those that will ensure many years of low maintenance. We follow this discussion with a brief exploration of homemade wind machines and we'll end with a brief discussion of windmills designed to pump water.

The Anatomy of a Wind Turbine

Small wind generators come in many shapes and sizes, but most belong to a group known as horizontal axis wind turbines or HAWTs. Most horizontal axis wind turbines are upwind

Alternator vs. Generator

A generator is a machine that converts mechanical energy into electrical energy, either AC or DC. A generator that produces AC is called an AC generator or, more commonly, an alternator. Alternators that produce DC electricity are known as DC generators or simply generators. Some classic small wind turbines, such as the Wincharger, contained DC generators and produced DC directly. Most, if not all, modern wind turbines contain alternators, which produce AC electricity. However, some wind turbines contain alternators that are equipped with rectifiers, devices that convert the AC to DC, which is then sent down the tower.

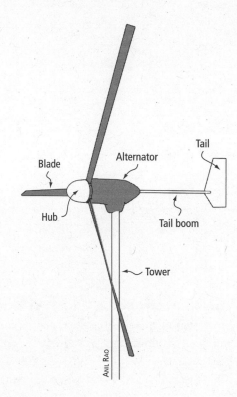

Fig. 5.1: *Anatomy of a Wind Generator. A wind turbine consists of blades attached to a hub. They form the rotor. The rotor in this drawing is attached via a shaft to the rotor of the alternator. When it spins, it produces electricity.*

main parts: (1) a rotor, (2) an alternator, and (3) a tail (Figure 5.1). Many modern turbines are equipped with three blades attached to a hub, covered by a nose cone. Together, the hub and blades form the rotor of the turbine (not to be confused with the rotor in the alternator). This entire assembly rotates when wind blows past the blades, hence the name "rotor." In some small wind turbines the rotor is attached to a shaft that's attached to an alternator, a device that produces AC electricity. In other wind turbines, the rotor is connected directly or through a gear box to the alternator. Alternators consist of two main parts, a set of stationary windings, known as the stator, and a set of rotating magnets, known as the rotor.

When winds cause the blades of a turbine to spin, the shaft spins. The spinning shaft, in turn, causes the rotor of the alternator to spin, creating electricity. Wind turbines, therefore, first convert the kinetic energy of the wind into mechanical energy (rotation). The mechanical energy is then converted into electrical energy in the alternator.

Figure 5.2 shows another common wind turbine configuration. In this drawing, you'll notice that there's no shaft. The blades of the turbine attach to a faceplate, which is attached directly to a cylindrical metal "can." Together the blades and the face plate form the rotor of the turbine. The can, to which the faceplate is attached, is the rotor of the alternator. It contains permanent magnets that spin around a set of stationary coils of copper wire, the windings. The windings constitute the stator of the alternator. As the rotor of the turbine spins, the

models — that is, the blades are located on the windward (upwind) side of the tower when the turbine is operating.

Among these turbines, you'll find two basic varieties: (1) turbines designed to charge batteries, appropriately called "battery-charging turbines," and (2) turbines designed to connect to the grid. They are referred to as "batteryless grid-tie turbines." Let's take a closer look at horizontal axis wind turbines.

Horizontal Axis Wind Turbines

Horizontal axis wind machines vary somewhat in design but all upwind turbines consist of three

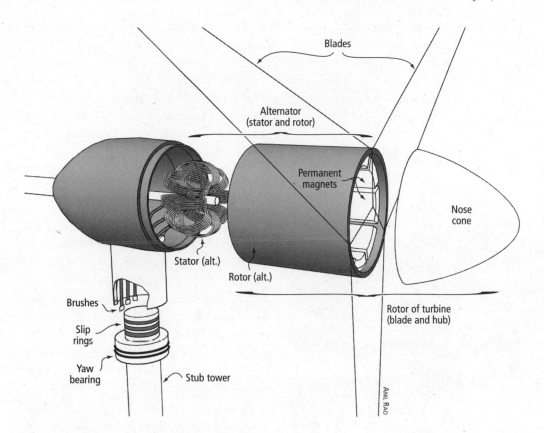

Blades

Alternator
(stator and rotor)

Permanent
magnets

Nose
cone

Stator (alt.)

Rotor (alt.)

Brushes

Slip
rings

Yaw
bearing

Stub tower

Rotor of turbine
(blade and hub)

ANIL RAO

Fig. 5.2: *Many small wind turbines now directly link the rotor of the turbine to the rotor of the alternator, as shown here. Note that the magnets are attached to the inside of the can. They rotate around the stationary windings of the alternator stator.*

magnets rotate around a stator, a set of stationary coils of copper wire (the windings). They make up the rest of the alternator. The rotation of magnets past the stationary (stator) windings produces alternating current electricity in the windings via electromagnetic induction. Electrical current is induced in the windings by the magnetic field from the rotating magnets.

Alternators in many small wind turbines contain rare earth magnets, rather than electromagnets which are found in the alternators of cars and trucks. (An electromagnet creates a magnetic field by running an electric current

through copper wire coils or windings). Rare earth magnets contain neodymium, iron and boron. They produce a much stronger magnetic field than conventional iron (ferrite) or ceramic magnets used in many small wind turbines. The stronger the magnetic field, the greater the output of a wind generator, all things being equal.

Generators that use magnets like rare earth magnets, as opposed to electromagnets, are referred to as permanent magnetic generators. The magnets are mounted on a rotor of the generator (the part that spins).

Electricity leaves the alternator via wires that attach to the stator. In most turbines, like the one shown in Figure 5.2, wires from the alternator terminate on metal brushes. The brushes, in turn, contact slip rings, brass rings located near the yaw bearing. The yaw bearing allows the turbine to turn in response to changes in the wind direction. The brushes therefore transfer electricity from the alternator to the slip rings. The slip rings, in turn, connect to a cable that runs down the length of the tower. Electricity therefore flows from the alternator to the brushes to the slip rings and then down the tower.

Fig: 5.3: *ARE110 wind turbine.*

Wind generators such as the ones shown in Figures 5.1 and 5.2 are known as direct drive turbines. That's because the rotor (blades and hub) are attached directly to the rotor of the alternator. As a consequence, the rotor and the alternator turn at the same speed. Although virtually all modern residential wind machines are direct drive, large commercial wind turbines and a few of the small wind turbines contain gearboxes that couple the rotor to the alternator. These are known as indirect drive or gear-driven turbines. (For more on this, see accompanying sidebar.)

Another important component of most horizontal axis wind machines is the tail. The

Direct Drive vs. Gear Drive

Large commercial generators and several small wind turbines are equipped with gearboxes interposed between the rotor and the generator. They increase the speed at which the alternator spins, thus increasing the output of the alternator. This, in turn, allows the alternator to be much smaller and also maintains a rotor speed (blade speed) that is safe and quiet. In the Jacobs 31-20 (a 20 kilowatt-hour-rated turbine with a 31-foot-diameter rotor), the gear ratio is 6:1. That means that the alternator turns six times faster than the rotor (the blades). Gearboxes are found in the new Endurance turbine and the remanufactured Jacobs turbines, such as the Jacobs 31-20. They are also used in the rebuilt Vestas V-15 and V-17 turbines, as well as in the Energie PGE and Entegrity turbines, all fairly large small wind turbines.

tail typically consists of a boom and vane. The tail boom connects the tail vane to the body of the turbine.

Tails are passive directional control devices. They keep the rotor of the wind turbine pointing into the wind. If the wind direction shifts, the tale vane turns the turbine into the wind, ensuring maximum electrical energy production. In the language of the wind industry, the rotation of a wind machine on a tower as it tracks the wind direction is referred to as yawing.

Although upwind turbines dominate the market, several manufacturers produce downwind turbines: Proven, Southwest Wind Power (Skystream), Entegrity, Ventera, and PGE.

Proven Energy, a Scottish company, for example, produces a line of durable, long-lasting downwind turbines, which are imported into North America. They are some of the most reliable turbines on the market. As shown in Figure 5.4, these wind turbines contain no tail. How does a tailless turbine orient itself into the wind?

In this and other downwind turbines downwind orientation occurs because the rotor is located several feet in front of and slightly to the side of the yaw bearing of the machine (located where the turbine attaches to the tower). When the wind blows, it can therefore push the rotor around to the downwind position. Yawing may also be controlled somewhat by the blades. The blades of the Proven turbines are hinged. When the wind is blowing and the blades are spinning, the hinged blades bend very slightly, forming a shallow cone. Coning, in turn, helps the rotor to orient downwind.

PROVEN

Fig. 5.4: *Proven downwind horizontal axis wind turbine.*

Downwind turbines work well. However, if the wind dies down and then reverses direction, they can get caught in the upwind position (with blades upwind from the tower). When stuck upwind, downwind turbines are unable to spin and generate electricity. Mick noticed this in 2007 soon after installing a brand new downwind turbine, a Skystream 3.7, on his property. The previous day, the wind had been coming from one direction, then died down in the afternoon. The next morning, the wind had shifted 180 degrees. Because the wind was blowing directly at the rotor, which was not properly oriented into the wind, the turbine could not reorient itself. Mick tugged on the guy cables to jiggle the machine, hoping that

The Dubious Promises of Vertical Axis Wind Turbines

Fig. 5.5: *Vertical Axis Wind Turbine.*
Although there's a lot of interest these days in vertical axis wind turbines, they are mounted at ground level, which exposes them to unproductive low-speed winds.

SANDIA NATIONAL LABORATORIES

As Mick writes in his small turbine column in the American Wind Energy Association's newsletter *Windletter,* "Yet another announcement about a technological breakthrough in the wind turbine industry has just arrived in my e-mail inbox. Invariably the device being unveiled is a vertical axis wind turbine ... Included with the excited chatter is an article or two published by some local newspaper," raising interest and false hope.

As shown in Figure 5.5, the blades of a vertical axis wind turbine (VAWT) are attached to a central vertical shaft, hence their name. When the blades of a VAWT spin, the shaft spins. The shaft is attached to an alternator generally located at the bottom of the shaft, sometimes even at ground level.

"Vertical axis wind energy devices have been around for a long time ... about 3000 years," notes Mick. These designs are so simple that many backyard tinkerers and garage inventors start with them, trying to fine-tune their ideas and equipment to make them work.

Proponents of VAWTs tout a number of supposed "advantages" over HAWTs that may, to the uninformed, seem to give this technology a decisive edge. One supposed advantage is that VAWTs can capture wind from any direction. The machines don't need to be oriented into the wind as the HAWTs do. Proponents also like to claim that VAWTs are immune to turbulence that wrecks havoc with HAWTs.

Another supposed advantage, say proponents, is that VAWTs don't need tall towers. They can be ☞

mounted close to the ground — even on top of buildings — where they capture ground-level winds. This, say supporters, eliminates the need for tall and costly towers and the need to obtain the zoning variances sometimes required to install horizontal axis wind turbines on tall towers.

Yet another supposed advantage of the vertical axis wind turbine stems from the fact that the generator can be mounted at ground level. This, say proponents, makes it easier to access and repair the generator should the need arise. There's no need to climb the tower — or lower a wind generator to the ground — to perform routine maintenance or to replace damaged or worn parts. Climbing towers or lowering them to the ground to maintain and repair a wind turbine takes time and can be risky to inexperienced individuals. Failure to inspect and maintain a machine can result in the early demise of a wind generator.

While these arguments cast VAWTs in a good light and are intended to convince us of their merit, most are invalid and others are greatly exaggerated. Moreover, years of experience with VAWTS has been rather discouraging, to say the least. "Hundreds of commercial VAWTs were installed in California in the late 1980s and early 1990s," Bob Aram reminds us. "They all failed and were removed from service. These were not experimental units, but production units."

Why have VAWTs failed to deliver on the promises of their supporters?

For one, the VAWTs are less efficient than horizontal axis wind machines. "For a given swept area," Jim Green notes, "they just don't extract as much wind energy as a well-designed HAWT." In addition, the blades of VAWTs are prone to fatigue created by centrifugal forces as the blades spin around the central axis. The vertically oriented blades used in some early models, for instance, twisted and bent as they rotated in the wind. This caused the blades to flex and crack. Over time, this caused the blades to break apart, leading to catastrophic failure. Because of this, VAWTs have proven less reliable than HAWTs and the early VAWTs have long since been abandoned.

"The VAWT does have an advantage in dealing with wind direction shifts," agrees Robert Preus, wind energy expert and manufacturer of Abundant Renewable Energy turbines (horizontal axis wind turbines), in his article, "Thoughts on HAWTs" in *Home Power* (Issue 104). However, rapidly changing wind direction that occurs in turbulent low-level winds increases fatigue on a VAWT, just like a HAWT. "Fatigue leads to equipment failure, which has been a major problem with VAWTs."

Many VAWTs also require large bearings at the top of the tower to permit rotation of the shaft. When the top bearings or the blades need replacement, you've got a job on your hands. Another problem with VAWTs is that many models require very thick and costly steel cable to support them. All of these contribute to the higher cost of electricity from VAWTs compared to HAWTs.

Next, although VAWTs can capture ground-level winds, just like any turbine installed on a ☞

too-short tower, they are just as sensitive to changing wind direction and turbulence as horizontal axis wind turbines. As you learned in Chapter 2, ground-level winds are subject to friction (ground drag). Low-level winds are also quite turbulent due to ground clutter. Both ground drag and turbulence in lower-level winds diminish the power available to a turbine when mounted at ground level — so much so, notes Mick, "that there is very little extractable energy in them." The lower the wind speed, the less electricity a turbine will produce. So just because a VAWT can be mounted at ground level doesn't mean that's a good idea.

While many modern VAWT inventors show videos of their turbines spinning, which convince news organizations of their viability, it's not spinning blades that matter. What matters is energy output. Because wind speeds are quite low at ground level, VAWTs can't produce much energy.

Don't forget, too, that eddies can form behind buildings and trees at ground level, resulting in dead air spaces, regions where the average wind speed is close to zero. Place a VAWT in a location such as this and you've just installed an expensive lawn ornament.

VAWTs are less reliable and efficient than HAWTs, and "these deficiencies are only made worse by mounting at ground level or on the top of buildings," notes Jim. What electricity they produce turns out to be more expensive as a result. All in all, they just don't stack up against horizontal axis wind turbines. The few advantages they offer cannot counter the many, some say fatal, disadvantages. ■

the turbine might reorient itself into the wind, but to no avail. He then went inside and the next morning found the turbine oriented correctly and working fine.

Getting stuck upwind is not unique to the Skystream, it's characteristic of all downwind passive yaw small turbines. However, most experts agree that this isn't a huge problem. It is more of an idiosyncrasy. As the wind speed increases or shifts direction, the turbine will align itself properly.

So far, we've seen that the most common wind generators are horizontal axis models. They can be direct drive or indirect drive. They can be upwind or downwind. There's also another type of wind machine that is getting a lot of attention these days. It is known as a vertical axis wind turbine. You can read about them in the accompanying feature box.

Improvements in Small Wind Turbines

Although small wind turbines in the 1970s and early 1980s had problems that gave the industry a black eye, market forces have eliminated many poor designs and marginal producers. Moreover, small wind turbines have improved dramatically. In fact, most small horizontal axis

---

Table 5.1
Trends in Wind Turbine and Wind System Design and Production

- Improved blade design and materials leading to greater strength and higher performance
- Greater generator efficiency, in large part by incorporating rare earth permanent magnets
- Reintroduction of induction generators that produce grid-compatible electricity without the use of an inverter
- Design for lower average wind speeds to improve output

- New and improved methods of overspeed control
- Reductions in rotor speed to reduce sound from blades
- Improvements in the efficiency and reliability of grid-connected inverters

Modified from Jim Green, Wind Energy Technology Center, National Renewable Energy Laboratory, Golden, Colorado

---

machines on the market today have been dramatically redesigned for simplicity, durability and high performance. Manufacturers have reduced or eliminated moving parts and parts that are subject to wear, such as brushes in alternators. (Brushes are still used in the yaw bearings to transfer electricity from the alternator to the wires leading down the tower.) They've substituted materials like rare earth magnets for ordinary magnets. And they've spent a considerable amount of time researching and testing blade materials and design to improve their efficiency. Table 5.1 summarizes many of the improvements in turbines.

Costly research and development, often carried out in conjunction with the US Department of Energy's Wind Energy Technology Center in Golden, Colorado, (part of DOE's National Renewable Energy Laboratory) has led to the production of much more durable small wind turbines, many of which can provide years of relatively trouble-free service. In fact, if properly sited, installed, and maintained, a high-quality

wind machine could easily last 20 to 30 years. That's no small feat, given the fact that a wind turbine in a good site operates as many hours in a year as an automobile operates over its lifetime. Improvements in the design and construction of small wind turbines have also resulted in longer warranties, typically five years.

Although improvements have resulted in much better wind turbines, buyers should not assume that all wind generators on the market today are the same. As you will see shortly, a number of less expensive light- and medium-duty turbines do not last as long as heavy-duty competitors. The predominant design principle in the lighter duty wind turbines seemed to be affordability. While cost is important, turbines designed principally with up-front cost in mind typically don't last long, especially if installed in high or turbulent wind sites or sites that experience a lot of storms with strong winds.

There are also a lot of cheap imports on the market that have not lived up to the manufacturers' expectations. Many of them failed to operate at all.

The Main Components of Wind Turbines

To help you select a wind turbine that's built to last, let's take a closer look at the main components of modern wind turbines, starting with blades and generators.

Blades

Blades are a vital component of a wind turbine. Manufacturers make blades from several different types of materials.

Materials

Although wood was once commonly used to make blades for small wind turbines, it has been replaced by plastics and composites such as fiberglass. In fact, we're not aware of any companies that currently manufacture blades exclusively from wood, except Otherpower

which produces blades for home made turbines. Wood blades are just too expensive and time consuming.

To ensure years of reliable service, the blades of modern wind turbines are typically made of extremely durable synthetic materials, various types of plastic or composites — plastic reinforced with fiberglass or carbon fibers, for example. These synthetic blades typically last between 10 and 20 years. Table 5.2 lists some examples.

Why use plastic?

Manufacturers have found that fiberglass and other plastics are not only cheaper than wood, they require a lot less labor to make. Unlike the metal blades once used in small wind turbines, plastic blades don't interfere with television, satellite TV or wireless Internet signals.

Two Blades or Three?

Most modern residential wind machines — both upwind and downwind models — have three blades. Although certain manufacturers produce two and six-blade turbines (for sailboats), three-blade wind turbines are the industry standard.

There's a reason for this — the number of blades used in a wind turbine is a compromise between efficiency, sound, longevity and cost. Let's begin with efficiency.

Interestingly, the most efficient number of blades is one. That is to say, a single-blade wind turbine is able to convert wind energy into electrical energy more efficiently than any other configuration. Unfortunately, single-blade designs

Table 5.2 Composition of Small Wind Turbine Blades	
Company	**Blade Material**
Abundant Renewable Energy	Fiberglass
Bergey Windpower	Fiberglass
Wind Turbine Industries	Fiberglass
Proven	Polypropylene (WT600) Fiberglass (WT2500, WT6000, and WT15000)
Southwest Windpower	Polypropylene reinforced with fiberglass Fiberglass

are difficult to balance when operating. As a result, you won't see single blade turbines in the product offerings of wind turbine manufacturers.

Two- and three-blade wind turbines operate at about the same efficiency (two blade turbines are slightly more efficient). All other factors being equal, two-blade turbines spin faster than three-blade turbines and produce slightly more electricity. In addition, two-blade turbines cost less to build than three-blade turbines, because they require less material and less labor. Because they spin faster, a manufacturer can use a smaller, less expensive generator. All this adds up to a less expensive turbine. If all this is true, then, why don't two-blade wind turbines dominate the market?

One reason is that three-blade wind turbines are quieter. Why?

One reason three-blade turbines produce less sound is that the more blades a turbine has, the lower the rotor speed. As a general rule, the lower the rotor speed, the quieter the turbine. (It should be pointed out that other factors like overspeed protection, discussed shortly, come into play when sound levels are concerned, but most of the sound generated by a small wind turbine is from the blades.)

Another reason three-blade turbines are quieter is that they don't produce yaw chatter, a phenomenon that plagues two-blade turbines as they turn in the wind.

Yaw chatter is a sound created by the vibration of the blades in a two-blade turbine as they move from a horizontal position (that is, perpendicular to the tower) to a vertical position (lined up with the tower) while the turbine

is yawing (turning). When perpendicular to the tower, inertia of the spinning blades poses maximum resistance to yawing. That is, inertia resists the turbine's attempts to pivot on the bearings as the turbine tracks changing wind direction. When the blades are in the vertical position, inertia is at a minimum and there's very little resistance to yawing. This rhythmic change — caused by the yawing motion that stops and starts during each revolution of the rotor — creates yaw chatter.

Chatter not only makes for a noisy turbine during yawing, it increases wear and tear on a turbine, which can lead to more maintenance and repair. It also considerably shortens the life of a turbine.

A three-blade rotor eliminates chatter because its rotational inertia is constant regardless of blade position. Although it costs more to manufacture, the quieter three-blade turbine is more popular among customers. It also suffers less wear and tear and lasts longer. Although initial cost may concern buyers, we urge prospective wind system owners to focus on the long-term cost.

Generators

Most residential wind machines on the market contain permanent magnet alternators, described earlier, although larger turbines in the small turbine range, like the remanufactured Jacobs 31-20, typically incorporate electromagnets. As noted in Chapter 3, most wind turbines produce wild three-phase AC, initially. In battery-charging systems, the wild AC output is converted (rectified) to DC electricity by a device known as a rectifier. The rectifier may be located in the wind turbine itself or in the charge controller. Charge controllers are typically mounted next to the inverter, which is located in or near the home or business. DC electricity is then sent to the battery bank.

As you may recall from Chapter 3, the charge controller regulates the flow of electricity to the batteries, preventing overcharge. If the batteries of an off-grid system are full, surplus electricity is fed into a diversion load, a resistance heater located in a water heater or a space heater. In grid-connected systems with battery backup, the surplus electricity is backfed onto the grid.

In off-grid systems or utility-connected systems with batteries, batteries deliver DC electricity to the inverter. The inverter converts DC to AC and boosts the voltage from 12, 24 or 48 volts, the voltage of the turbine and battery bank, to 120- or 240 volts, the voltages required in most households and businesses.

In utility-connected systems, wild AC from the turbine is also converted (rectified) to DC in the controller or inverter. However, it is sent to a special grid-connected inverter in systems containing separate controllers. The inverter converts the DC back to AC but a much "tamer" version, one that matches grid power — 60 cycle per second 120- or 240-volt electricity in North America (in synch with the utility's grid electricity). The electricity is then delivered to active loads in the house via the main service panel. The excess is backfed onto the electric grid.

Also noted briefly in Chapter 3, some manufacturers are now producing small turbines with induction generators. These wind turbines produce grid-compatible AC power without the use of an inverter. An induction generator looks a lot like electric induction motors widely used in modern society. When spun faster than its normal operating speed, an induction generator produces AC electricity. Because an induction generator synchronizes with the grid — that is, produces electricity at the same frequency and voltage — no inverter is required. (However, other electronic controls are needed.)

With these basics in mind, we now turn our attention to specific features to look for when buying a wind machine.

What to Look for When Buying a Wind Machine

If wind turns out to be a viable option for you, you'll need to select a wind turbine. While there are many turbines on the market, careful load and site analysis should have narrowed the field considerably. Once you have determined your average monthly electrical load and the average wind speed on your site, you can select a wind turbine that will produce enough electricity to meet your demands. For example, if the wind site analysis showed that the average monthly wind speed at your site was 13 miles per hour and your load analysis showed that you need around 450 kilowatt-hours a month or 5,400 kilowatt-hours per year, after implementing efficiency measures, and if you were planning on installing a utility-connected system, you'd have only three or four choices of turbines, as indicated in Table 4.4 on page 98: (1) a Proven WT2500, which produces, on average, about 645 kilowatt-hours per month, according to the manufacturer; (2) an ARE 110, which produces, on average, about 505 kilowatt-hours per month; (3) a Skystream 3.7, which produces approximately 450 kilowatt-hours per month; and (4) perhaps a Whisper 500, which produces 620 kilowatt-hours per month. But how do you choose among these options?

Manufacturers provide a plethora of technical data on their wind machines that can be used to make comparisons. Unfortunately, most of it is useless. Further complicating matters, "There can be a big difference in reliability, ruggedness, and life expectancy from one brand to the next," according to Mike Bergey.

So how do you go about selecting a wind machine?

Although wind turbines can be compared by many criteria, there are only a handful that really matter: (1) swept area, (2) durability, (3) annual energy output, (4) cost, (5) governing mechanism, (6) shut-down mechanism, and (7) sound. Let's take a look at each one.

Swept Area

In wind energy, rotor size is everything — well, almost everything. To get the most out of a wind turbine — to produce the most electricity at the lowest cost — we recommend selecting a wind turbine with the greatest swept area.

Swept area is the area of the circle described by the spinning blades of a turbine. Because the blades collect energy from the wind, the swept area is the collector area of a turbine. It is akin to the square footage of a solar-electric array. Doubling the area of a solar-electric array (the area exposed to the sun) doubles its output. The same holds true for wind machines. As a rule of thumb, then, the bigger the swept area, the more energy you'll be able to capture from the wind. Swept area allows for easy comparison of different models. Look for the models that have the greatest swept area. As Hugh Piggot of Scoraig Wind Electric in Scotland puts it, "Swept area is easier to measure and harder to lie about than performance."

Swept area is determined by rotor diameter. When comparing wind turbines, then, the rotor diameter is a pretty good measure of how much electricity a turbine will generate. Although other features such as the efficiency of the generator and the design of the blades influence energy production, for most turbines they pale in comparison to the influence of rotor diameter and swept area.

Manufacturers list the rotor diameter in feet or meters — often both. The rotor diameter is the distance from one side of the circle created by the spinning blades to a point on the opposite side or about twice the length of the blades. The greater the blade length, the greater the rotor diameter and the greater the swept area — that is, the greater the amount of wind a turning rotor intercepts.

Fortunately, most manufacturers also list the swept area of the rotor. Swept area is presented in square feet or square meters — sometimes both. Table 5.3 lists the rotor diameter and swept area of the four residential-sized wind machines used in our example.

To illustrate the importance of swept area, take a look at Figure 5.6. It illustrates the

Table 5.3
Comparison of Four Wind Generators*

Wind Turbine	Rotor Diameter (ft)	Swept Area (ft²)	Weight (pounds)	Annual Energy Output (AEO) at 8 mph**	AEO at 10 mph**	Cost (10/2007)	Cost per square foot swept area	Cost per pound	Value of Electricity at 12 cents per kWh/ year/ Return on Investment
Proven WT2500	11.1	97	419	2004	3516	$10,140	$105	$24	$422/4.2%
Are110	11.8	110	315	1620	3144	$11,500	$105	$37	$377/3.3%
Skystream	12	113	170	1200	2880	$5,400	$48	$32	346/6.4%
Whisper 500	15	176	155	2040	3960	$12,125	$69	$78	$475/3.9%

*This table does not include tower costs, which must be taken into account when making a decision.
**Based on manufacturer's estimates.

NOTE: Shading is meant to indicate which turbines are best based on each criterion (in each column)

JOHN RICHTER

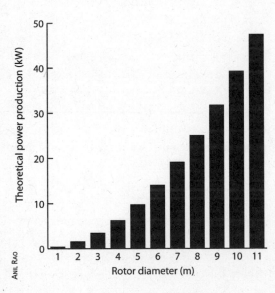

Fig. 5.6: Electrical Production vs. Rotor Diameter. Theoretical power production of a small wind turbine at wind speed of 22.4 miles per hour (10 meters per second). Notice the effect that increasing rotor diameter has on power production.

ANIL RAO

theoretical increase in power production in a 22.4 mile per hour (10 meter per second) wind. As you can see, power production theoretically increases from 0.4 kilowatt-hours to 1.6 kilowatt-hours with a doubling of rotor diameter — from 1 meter to 2 meters. When rotor diameter increases from 2 to 4 meters, theoretical power output increases from 1.6 kilowatt-hours to 6.4 kilowatt-hours. (Remember that these are theoretical estimates. Other factors such as turbine and blade efficiency and air density influence electrical output.)

Annual Energy Output

Although greater swept area generally translates into greater electrical production, another, even

Converting Rotor Diameter to Swept Area

As noted in the text, some manufacturers list rotor diameter rather than swept area in their promotional literature and spec sheets. Because small differences in rotor diameter translate into substantial differences in swept area, it is always worth calculating the swept area of the wind turbines when you are comparing one model to another.

To calculate swept area from rotor diameter, start with the rotor diameter. Divide diameter by two to determine radius of the swept area, which is usually close to the blade length.[1] (The radius of a circle is half the diameter.) Now plug the radius into the equation for the area of a circle: Area $= \pi r^2$. As you may also recall from middle school math, the symbol π is pi, a constant, 3.14.

To see how a small increase in blade length or rotor diameter adds up to huge increases in swept area, let's compare the swept area of two turbines, one with a rotor diameter of 10 feet and another with a rotor diameter of 11 feet.

In the wind turbine with a ten-foot rotor, the radius of the swept area is five feet. The swept area is 3.14 x 5^2 = 3.14 x 25 = 78.5 square feet. In the turbine with a rotor diameter of 11 feet, the radius of the swept area is 5.5 feet; however, the swept area is 95 square feet (3.14 x 5.5^2 = 95). In this example, a 10 percent (0.5 foot) increase in radius resulted in a 21 percent increase in swept area.

more useful, measure is the annual energy output (AEO) at various wind speeds. Developed by the American Wind Energy Association (AWEA) to help standardize rating systems for wind turbines, the AEO of a given wind turbine is presented as kilowatt-hours of electricity produced at various average wind speeds. Like the US EPA's estimated gas mileage for vehicles, AEO gives buyers a convenient way to compare models. As in the estimated gas mileage rating, however, AEOs won't tell you exactly how much electricity a wind machine will produce at a site. Performance varies, as it does in cars and trucks. In cars and trucks, gas mileage is affected by the way an owner drives the vehicle and the air pressure in the tires.

With wind turbines, AEO is affected by tower height, turbulence and the density of the air, among other factors.

In addition, AEO is calculated by the manufacturers. Some are conservative in their calculations so as not to disappoint owners with exaggerated estimates of energy production. Other manufacturers are more "optimistic" in their energy estimates, but their products may fail to deliver.

Despite these caveats, AEO provides a useful means of comparison. As shown in Table 5.3, the Proven WT2500 should produce approximately 2,004 kilowatt-hours per year at sites with an average wind speed of 8 miles per hour and about 3,516 kilowatt-hours per year at sites with an average wind speed of 10 miles per hour, according to the manufacturer's estimates. The ARE 110 should produce approximately 1,620 kilowatt-hours per year in a site with an average 8 mile-per-hour wind and 3,144 at 10 miles per hour, again according to manufacturer's estimates. The Skystream 3.7 should produce approximately 1,200 kilowatt-hours at 8 miles per hour and 2,880 at 10 miles per hour. The Whisper 500 should produce approximately 2040 kilowatt-hours at 8 miles per hour and 3,960 at 10 miles per hour.

Durability: Maximum Design Wind Speed and Tower Top Weight

Another extremely important criterion to consider when comparing wind turbines is durability. One measure of durability you'll see in advertisements is the maximum design wind speed. Maximum design wind speed is the wind speed a machine can supposedly withstand without sustaining damage. At first glance, you'd think that this might be a good parameter by which to compare turbines. It stands to reason that a wind turbine designed to withstand 130 mile-per-hour winds ought to be more rugged than one designed to withstand 120 mile-per-hour winds.

The problem with maximum design wind speed is that, while wind generators are designed to survive wind speeds of 120 mph or more, they are either not tested at these speeds or are not repeatedly tested at these speeds by manufacturers. Another problem with maximum design wind speed is that wind generators are more often damaged by turbulence rather than

by high-speed winds. Moreover, even in high winds, many individuals in the industry contend that it is not the wind that damages a wind generator, but flying debris, although Ian disagrees. He has never witnessed damage attributable to flying debris, but has seen plenty of damage caused by high winds.

A far better measure of durability is tower top weight — how much a wind turbine weighs. Let's examine the tower top weight of the four wind turbines we've been considering. As shown in Table 5.3, the Proven WT2500 weights 419 pounds, the ARE 110 weighs 315 pounds, the Skystream 3.7 weighs 170 pounds, and the Whisper 500 weighs 155 pounds.

In our experience, heavyweight wind turbines tend to survive the longest — sometimes many years longer than medium- or light-weight turbines. Be forewarned, however: weight is usually reflected in the price. As a rule, the heavier a wind machine, the more you'll pay. Remember, however, that you get what you pay for. Producing electricity on a precarious perch 80 to 165 feet above the ground isn't a job you want to relegate to the lowest bidder, which is invariably the lightest turbine.

Cost

Although swept area, annual energy output (AEO), and tower top weight are the primary criteria one should consider when shopping for a wind machine, cost is also important. For many potential small wind energy producers, initial cost is often used to compare wind turbines. However, there's a better way to compare the cost of wind turbines than by simply examining initial cost. That is, to compare costs based on the cost per unit of weight and cost per unit of swept area.

Table 5.3 lists the cost of the four wind turbines we have been using in our example. Based on cost per unit of swept area, the Skystream 3.7 appears to be the best bargain. Based on cost per pound, the Proven WT2500 comes out on top.

Another way to compare wind turbines is to calculate their simple return on investment. In Table 5.3, we've included the annual value of the electricity generated by the turbines at 12 cents per kilowatt-hour, on an average wind speed of 10 mph. We've then calculated a simple return on investment for each turbine, a topic discussed in Chapter 4. Based on these calculations, the Skystream and Proven come out number one and two once again.

Now that you've analyzed your options, it's time to make a decision. Which turbine is the best buy based on the criteria shown in Table 5.3?

As you can see from that table, the choice is not always clearcut. The two lighter-weight turbines fare pretty well. However, even though a lightweight turbine may appear to be a bargain, if it lasts half as long as a heavier machine, you'll end up paying more in the long run. You may end up buying two or three of them over a 20-year period. When calculating cost or shopping for a turbine, then, be sure to take into account durability and estimate costs based on lifespan. Although we can't quote exact lifespans, and any attempt to do so would be suspect, it is a pretty safe bet that the lighter the wind turbine, the lower its lifespan.

It's also a good idea to add in the balance of system cost (see sidebar) and consider the credibility of the performance data you've been studying. Was it produced by the manufacturer and verified by an independent third party?

When examining test data, be wary of results from wind tunnel studies or truck testing — both are notoriously inaccurate. Also note that some claims of increased efficiency may be valid in a laboratory, but not in the real world. For example, a hot new airfoil that delivers impressive power fresh out of the box may not do so in the real world when the paint finish has matted after a few months of weathering, or dirt or dead insects build up on the leading edge of the blade. Increasing the rotor diameter an inch or two may result in better performance than an "improved" airfoil.

Balance of System Cost

Before you buy a machine, it's also a good idea to consider total system cost. You'll need to purchase a tower and pay for installation, unless, of course, you install the turbine and tower yourself. Even then, you'll need to pay for concrete, rebar and equipment to excavate the foundation and anchors. You'll also need to run electrical wire from the turbine to the house and purchase an inverter (although they're included in most batteryless grid-tie wind turbines). If you're going off-grid or want battery back up for your grid-connected systems, you'll also need to buy batteries. All of this will add to the cost. Cost of the turbine may range from 10 to 40 percent of the total system cost, a topic covered in Chapter 10.

When looking for information on wind turbine performance, seek out performance data derived from field tests — that is, studies of turbines on towers exposed to the elements. One source of field test data is the National Renewable Energy Lab's Wind Energy Technology Center. They post data on several small wind turbines on their website, including sound production data, and will be adding more data thanks to a new small wind turbine test program.

Another source of data is Appalachian State University's Small Wind Research and Demonstration Facility on Beech Mountain in North Carolina. They test small wind turbines under some fairly extreme wind conditions. You can also find data from various end users on the Internet.

Although you now know a lot about wind turbines, there are a few other factors that you should take into consideration when buying a wind turbine, including the governing mechanism.

Fig. 5.7: Many microturbines like the Marlec (shown here) have no governing mechanism to slow the rotor in high winds. They rely on the relatively low rotor speed and rugged construction to endure high winds.

MARLEC

Governing Systems

Found in all wind generators worth buying, governing, or overspeed control, systems are designed to prevent a wind generator from burning out or breaking apart in high winds. They do this by slowing down the rotor when the wind reaches a certain speed, known as the governing wind speed. Why is this necessary?

As wind speed increases, the rotor of a wind turbine spins more rapidly. The increase in the revolutions per minute (rpm) increases electrical output. Although electrical output is a desirable goal, if it exceeds the machine's rated output, the generator could overheat and burn out. In addition, centrifugal forces in high wind speeds exert incredible forces on wind turbines that can tear them apart if the rpm is not limited. Designing a wind turbine to handle all that force would make an extremely heavy machine that would likely not perform well in moderate winds found in most locations, the winds we capture to produce most of our electricity.

A governing system is essential because it allows the turbine to shed extra energy when the winds are really strong and survive so it is available for the next capturable wind. Not all wind turbines come with governing mechanisms, however. Many of the smallest wind turbines, the microturbines, with rated outputs of around 400 watts, for example, have no governing mechanisms (Figure 5.7). (These turbines are too small to produce a significant amount of electricity for applications covered in this book.) Larger wind turbines, those with swept areas over 38 square feet, however, come with overspeed controls. Two types are commonly found: furling and blade pitch.

Furling

Most manufacturers protect their wind turbines by furling. Furling is an ancient term that originated in the days when windmills were built with fabric-coated blades. The miller who operated the wind machines climbed onto the blades and rolled up or furled the fabric so they would no longer turn. This protected the windmills from damage in high winds.

Today, the term is used to describe a very different process that achieves the same end result. Furling is accomplished in one of two ways, both of which shift the position of the rotor (hub and blades) relative to the wind, effectively turning the blades out of the wind. This, in turn, decreases the amount of rotor swept area (known as frontal area) that intercepts the wind. Reducing the swept area reduces the speed at which the rotor turns and the energy collected. Slowing the rotor protects the wind turbine from damage — or at least that's the theory.

Manufacturers employ two main types of furling: horizontal and vertical. In horizontal furling, the rotor turns out of the wind by turning sideways. For this reason, horizontal furling is also known as side furling. In vertical furling, the rotor rotates upward with the same effect. "Angle furling" is a combination of the two.

Horizontal or side furling is achieved, in part, by hinging the tail. In side-furling turbines, a hinge is located between the tail boom and the body of the turbine. As you can see

MICK SAGRILLO

from Figure 5.8, the turbine is also slightly offset from the yaw axis — that is, the yaw bearing is attached to the side of the turbine body, not its center so the turbine is not directly over the tower.

Because the turbine is offset from the yaw axis, the force of the wind on the blades tends to rotate the machine around the yaw axis. However, the tail resists this rotation and keeps the rotor facing into the wind.

In light winds, the forces on the rotor and tail are small and the wind holds the tail in its

Fig. 5.8 : *Side Furling. This wind turbine is not broken, it is side furling in high winds, which slows the rotor and protects the machine from damage.*

Fig. 5.9: *Vertical furling.*

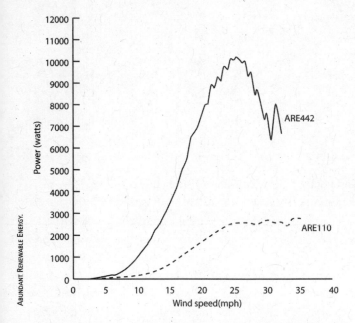

Fig. 5.10: *Power Curves. The electrical production of the ARE442, like that of many other turbines, declines in high wind speeds as a result of overspeed controls that protect the wind turbine from damage. The power production of the ARE110 pleateaus.*

normal position — straight behind the turbine. However, in strong winds, the increasing forces on the rotor overcome the force of the wind on the tail. Since the tail creates more force than the offset rotor, the tail stays mostly aligned with the wind and the turbine turns away from the wind. As a result, the turbine literally folds on itself. This reduces the frontal area of the rotor, which slows the rotor.

Vertical furling is achieved by moving the hinge in front of the yaw axis. In high winds, the force of the wind tilts the rotor up, while the tail stays oriented downwind. As in side furling, this reduces the frontal area of the rotor and reduces its speed. The lifting force of the wind is opposed by the weight of the rotor and generator. (Figure 5.9).[2]

When fully furled, the rotor of a vertical furling turbine resembles a helicopter rotor. When wind speed declines, however, the rotor returns to its normal operating position. Shock absorbers are often used to ease the rotor back into position.

Furling reduces the amount of energy collected by the rotor. Although electrical output may continue, it typically occurs at a lower rate, as shown in the power curves in Figure 5.10. When the wind speed diminishes, however, the turbine unfolds, returning to its normal operating position.

Changing Blade Pitch

The second type of overspeed control involves a change in blade pitch — that is, a change in the angle of the blades — to reduce rotor speed. Blade pitch changes are used in propeller-driven airplanes to control power.[3] It works the same way in wind turbines. When wind speed increases to a dangerous level — the upper limit of the operating range of the machine — blade pitch changes automatically.

Changing the angle of the blade reduces its aerodynamic lift. Reducing lift, in turn, reduces rotor speed. The greater the wind speed above the operating range of the machine, the more the blades rotate (pitch). The greater the change in pitch, the lower the lift.

Blade pitch control is found in a number of small wind turbines, among them Kestrel wind turbines, manufactured by a subsidiary of Eveready, Ltd., the South African company that bring us batteries. Blade pitch turbines are also manufactured by Proven, Jacobs and Eoltec.

Pitch control typically requires springs, gears and weights ingeniously arranged to produce the desired effect (Figure 5.11). Some machines like the Jacobs use the weight of the blades to change the pitch.

While many manufacturers incorporate one type of governor, Abundant Renewable Energy which manufacturers the ARE110 and ARE442, combines two mechanisms to achieve this end. It uses side furling and a form of dynamic braking for overspeed control (dynamic braking is described in the text). ARE turbines side furl at high wind speeds, but also direct some power to a resistor bank. The resistor bank increases the load on the generator which reduces rotor speed.

Fig. 5.11: *Numerous ingenious methods of blade pitch control have been devised. In this turbine, a Jacobs 31-20, the springs are part of a complex and effective blade pitch control mechanism.*

Dan Chiras

Blade pitch functions admirably, but is not as widely used as horizontal and vertical furling mechanisms. Why?

For many small wind machine manufacturers, pitch control has proven too costly. Some even argue that pitch control mechanisms are less reliable and require more maintenance than the average homeowner is inclined to perform.

As in many issues, universal agreement on this subject is lacking. Many people who own and operate turbines with blade pitch control argue that they operate better than furling.

Mick sides with proponents of blade pitch governors. Even though there are more moving parts, he prefers blade-pitch governors over furling anytime! Although furling mechanisms may be cheaper, Mick considers furling to be less reliable. "Furling may or may not work," he says. In addition, most furling mechanisms in use today don't slow the blades as they are supposed to and therefore don't provide the protection they're supposed to offer. Why have a governor if it's unreliable?

On the issue of maintenance, Mick does admit that blade pitch mechanisms need to be rebuilt every 10 to 15 years. However, furling governors require similar maintenance in the same time frame. In these machines, the bearings and bushings that are part of the passive furling mechanism must also be replaced every 10 to 15 years, although they are cheaper and easier to replace or rebuild than a pitch control system.

Both furling and blade pitch governors control rotor speed in high winds. However, blade pitch governors seem to give better control of speed and may be more reliable than furling. Bottom line: although furling mechanisms are cheaper, cheap is not necessarily better when it comes to a wind machine. The goal in buying a wind machine is not to get the cheapest turbine, but to purchase the most reliable and most durable turbine you can afford. Reliability and durability beat cheaper up-front cost any day.

While the governing mechanism is an important consideration, you may not have a choice, because you may only have a few choices among the turbines that produce the amount of electricity you need and most of the home-scale wind turbines use furling. In the example we've been using, however, that's not the case: the Proven WT2500 is governed by blade pitch; the ARE110 is governed by side furling and a dynamic brake; the Skystream has a dynamic brake; and the Whisper 500 is governed by angle furling.

Shut-Down Mechanisms

Because wind machines require routine maintenance and occasional repair, and because these operations may need to take place on slightly windy days — or winds may come up on calm days — small wind turbines should include a reliable shut-down mechanism. As the name implies, shut-down mechanisms allow a turbine to be turned off, permitting operators to maintain and repair wind turbines without fear of injury. They also provide a means of shutting a wind machine down when extremely violent storms, especially thunderstorms, are approaching. In addition, they allow operators to shut down wind turbines when leaving town for a while if they're concerned that a violent thunderstorm may strike while they're gone.

Even though all wind turbines worth considering come equipped with overspeed controls, which should protect a wind turbine in high winds, they rarely have a backup brake, unlike commercial wind turbines. (Commercial wind turbines include disc brakes that hold the rotor in any wind, but are also equipped with a safety plug — a stick brake — that is inserted through the rotor shaft to block its rotation. Thus, they have two redundant safety mechanisms that are both engaged before a worker gets near the rotor.)

Because small wind turbines typically lack a backup brake, small wind turbines can't be reliably secured in storms. They also can't be reliably secured when someone's working on a turbine. (As a result, maintenance personnel engage the shut-down mechanisms, but also secure the blades with rope in case the wind comes up while they're servicing a turbine on a tower.)

Wind turbine designers rely on two types of shut-down mechanisms: mechanical and electrical. Let's start with mechanical systems.

Mechanical systems include disc brakes and folding tails. Disc brakes, like those found in our cars and on some more expensive bicycles, are manually activated. They're attached to a cable that runs down the tower. Tightening the cable activates the brake and stops the rotor from turning. Wind generators from Proven Energy and Wind Turbine Industries (Jacobs)

Hurricane Protection

Even though some wind turbines have survived hurricanes, it's best to lower the turbine in a hurricane to protect it from flying debris if you have a tilt-up tower. Braking a turbine on a fixed tower will reduce the chances of damage.

incorporate cable-operated disc brakes. In these turbines, "brakes off" is the default setting. In other words, you have to tighten the cable to activate the brakes.

Although disc brakes like this may seem like a good idea, they are not fail-safe. If the cable breaks in violent storm, for example, an operator would be helpless to stop the turbine. There's no way to apply the brakes!

A far better design is one in which "brakes on" is the default setting. In these turbines, the operator has to release the brake to permit the turbine to operate. If the cable breaks, however, the disc brake is activated and the wind machine is brought to a halt.

While "brakes off" is the default of most wind turbines, there are some exceptions. One of them is the Endurance induction wind turbine. It incorporates a fail-safe design that relies on a spring-actuated brake. (This brake is applied by a strong spring and is released by air pressure controlled by an electric valve.) This turbine can be shut down automatically by its controller, and if the electrical system or air system fail, the brake is applied automatically by the spring.

Fig. 5.12:
Clay Sterling, MREA's Education Director, shuts down a Jacobs wind turbine by tightening the cable attached to the tail of the turbine.

Some small wind turbines come with tail vanes that can be manually folded via a cable attached to the base of the tower. When the tail is folded, the rotor is turned edgewise to the wind, greatly reducing its ability to collect energy. As in the example shown in Figure 5-12, a cable winch is located at the base of the tower. It allows the operator to manually shut down a wind machine in fierce winds or when leaving town for a while.

Although folding the tail protects the rotor from overspeeding, it doesn't stop it from rotating. This presents a potential risk to service personnel working on the tower, unless another means of stopping the rotor, such as a

disc brake or dynamic braking, is available. Furthermore, if the cable breaks in high winds, when the machine is shut down, the tail will swing back into the wind and the wind turbine will start back up, which could damage the turbine if the winds are strong enough.

Wind turbine designers also employ electrical brakes, known as dynamic brakes. Dynamic braking is the least expensive option and is found in most small-scale wind turbines.

Dynamic braking is a fairly simple approach that's unique to wind turbines that use permanent magnet alternators. It consists of a switch inside the house or at the base of the tower. When the brake switch is closed, it short circuits the wind machine, rapidly slowing the rotor. Dynamic brakes short circuit the three phases of the permanent magnet alternator. Electricity then flows through the windings of the stator. This produces a magnetic field around the windings that opposes the magnetic field of the rotor, slowing, even stopping the turbine. That is, it overpowers the ability of the rotor to spin the alternator.

In dynamic braking, the braking force is proportional to the rotor speed. As the rotor slows down, the braking force diminishes. As the rotor speed approaches zero, so does the braking force. In low to moderate winds, dynamic braking should either stop the rotor or slow it down considerably. However, dynamic braking may not completely stop the rotor in high winds. In winds blowing over 20 miles per hour.

Mick's found that dynamic brakes can't be counted on. Other individuals confirm his experience. Therefore, if a wind machine is shut down prior to a storm's arrival, strong winds may overpower the brakes, causing the rotors to start turning. In high wind speeds that force the blades to start spinning slowly, says Mick, energy is dissipated in the windings of the alternator, which could cause it to burn up.

Although dynamic brakes of some manufacturers are not 100 percent reliable, some are. Mick has found that the dynamic brakes of Southwest Windpower's Skystream 3.7 and turbines made by Proven and Abunadant Renewable Energy work well.

Shut-down mechanisms of a wind turbine should be high on the list of considerations, right up there with swept area and tower top height. If the turbine is to be serviced on the tower, the shut-down mechanism should be capable of completely stopping the rotor.

Unfortunately, some small wind machines (less than ten-foot diameter rotors) come without any shut-down mechanism. Like governing mechanisms, they are expensive, so some manufacturers leave them out. While this makes their machines cheaper initially, they are vulnerable to catastrophic failure. It goes without saying that replacing a wind turbine after its first thunderstorm is much more costly than purchasing and installing a machine with a reliable shut-down system.

Without a shut-down mechanism homeowners are left to chance. A violent storm may destroy their turbine and they're powerless to do anything about it except lower the machine to the ground, if possible. (That's not a good idea in a strong wind!) Mick's heard many horror stories of small wind machines that were destroyed in strong winds while the homeowners looked on helplessly. Most of them had no idea that their inexpensive wind turbine could be irreparably damaged in a storm until after the fact.

"Inexpensive wind turbine designs without a reliable shut-down mechanism are a short-sighted gamble at best."

— Mick Sagrillo

Sound Levels

The sound a turbine produces is also a factor one should consider when buying a turbine. Sound levels are important to homeowners and their neighbors. They may be important to zoning officials, too, as you shall see in Chapter 10.

All residential wind machines produce sound. There's no such thing as a perfectly quiet wind turbine, although some get quite close.

Sounds emanate from four possible sources. When the blades spin they produce a swishing sound. When furling, the blades may produce a helicopter-like beating sound. (This occurs when a blade passes through the wake of a preceding blade.) Rotation of the rotor in the alternator also produces sound, depending on the controller design. In gear-driven wind turbines, gears also produce sound. (Sound test reports can be found at the National Renewable Energy Laboratory's website, nrel.gov/, search term "sound test reports".) "Another source of sound from a wind turbine is generated in the alternator of some wind machines. It is caused by the interaction of the alternator and the inverter," says Jim. "The high switching frequencies in the inverter (at several kiloherz) distort the current and the voltage in the alternator. This, in turn, vibrates the stator and generates sound."

Sound levels increase as wind speed increases. However, sound from a wind turbine is often difficult to detect and is rarely a nuisance one could label as noise. Why?

As we have pointed out numerous times in this book, wind turbines are typically mounted fairly high off the ground — typically 80 to 120+ feet — to reach the smoothest, most powerful winds. Because sound decreases with the square of distance, tower height significantly reduces sound levels on the ground. In addition, sound from wind machines is often drowned out by environmental sounds. That's because, background sounds increase as wind speed increases, too. When the wind blows, leaves and branches of trees rattle and shake. Wind blowing by one's ears also makes sound. All of this background noise tends to drown out sound produced by a wind generator.

Even so, it is important to consider sound levels. One way is to observe turbines you are considering in operation under a variety of wind speeds. If you can't, you may want to ask homeowners or business owners who have installed the turbines you are considering for their experiences.

Another, more scientific, method is to check out the rpm of the turbines you are considering at their rated outputs. Rated output, discussed shortly, is the output of a machine in watts or kilowatts at a certain wind speed. Knowing how fast the blades and generator are rotating at that speed gives you an idea of the potential for sound. Generally, the higher the rpm, the more sound the blades will make. Bear in mind, though, that it is the tip speed, not just the rpm, that affects the amplitude of the sound. A small turbine at 1,000 rpm could be quiet and a much larger turbine at 50 rpm could be very loud.

In the turbines we've been considering, the Proven, ARE and the Skystream rotate at 300, 340 and 325 rpm, respectively. The Whisper rotates at 900 to 1,000 rpm. All of these machines, except the Whisper 500, should be pretty quiet. Mick's also found that the Skystream makes sound because the blades are downwind from and so close to the tower.

While the rpm of a wind turbine influence sound levels, they also give an indication of quality. Generally, less expensive and less durable turbines spin at a higher rpm. They rely on less expensive generators operating at high speeds to produce energy. In addition, higher rpm machines are subject to more wear and tear and tend not to last as long.

Other Considerations

Although swept area, weight, annual energy output, governing mechanisms, shut-down mechanisms, and sound levels are the most important factors to consider when buying a wind machine, there are other details that manufacturers provide. We think readers should be familiar with them, but not let them overly influence their judgment. They're nowhere near as important in the final analysis.

Cut-In Speed

One factor that is frequently advertised is cut-in speed — the wind speed at which a wind generator starts producing electricity. Although the blades of most small wind machines start turning in low winds, most turbines don't produce measurable amounts of electricity until wind speeds reach six to eight miles per hour. They produce full power at speeds from about 23 to 30 miles per hour.

Some manufacturers recommend installing a turbine with the lowest cut-in speed possible in areas of low wind to make better use of the wind resource. However, because low wind speeds hold very little energy, low speed cut-in isn't overly important. As an example, let's compare the output of a turbine in a six mile-per-hour wind to a wind turbine in a ten mile-per-hour wind. Knowing that power is a function of velocity cubed, we can compare output by cubing six (don't worry about units) to cubing ten. When we compare 6 x 6 x 6 (216) to 10 x 10 x 10 (1,000), you can see why low-wind-speed cut-ins are irrelevant. Winds in the 10 to 20 mile per hour range are the ones we want to capture. In the course of a year, wind speeds between 10 and 20 miles per hour deliver the bulk of the annual energy output. Wind speeds below ten miles per hour contribute only marginally. Any company that uses cut-in speed or very low wind speed performance as an advertising point does not understand how wind works.

Peak Output

Another relatively useless bit of information that manufacturers provide on spec sheets and in advertisements is peak output. Peak output is the maximum number of watts a wind turbine can produce — usually slightly above the rated wind speed. While this figure may be of interest to engineers, it is of little relevance to buyers. The question you should be asking, according to wind energy engineer Eric Eggleston, is: "What will this wind generator do at my site in my average wind speed?" In other words, what's the AEO, the annual energy output?

Power Curves

Another intriguing but ultimately useless bit of information provided by wind turbine manufacturers is the power curve, an examples of which is shown in Figure 5.9. Power curves are graphs that show the power production of a wind turbine in watts at different wind speeds.

While power curves are fascinating and make for good visuals, they aren't always accurate. Aaron Godwin of The Renaissance Group in Ohio, which installs wind systems, notes that "power curves put out by manufacturers are wishful at best." Moreover, they are difficult for most people to interpret. When shopping for a turbine, what's important is how much energy a wind turbine will produce at your site on your tower at your average wind speed — that is, AEO.

Rated Power

Last but not least is a measurement that's about as useful as payback, discussed in Chapter 4. It's known as rated power, a figure we've avoided as much as possible in this book.

Rated power is the output in watts at rated wind speed (defined shortly). It's often slightly lower than the peak power, the maximum output of a turbine.

Much like the output of solar electric modules, rated power of wind turbines was devised to give customers a way to compare products. Customers, for instance, typically compare one kilowatt or ten kilowatt turbines as they would compare solar modules.

Although rated power may seem like a useful tool, its usefulness is limited, in large part, by the fact that there are currently no standards in the wind industry (as there are in the solar industry) for determining rated power. All manufacturers determine rated power at different wind speeds, known as the rated wind speed. Hopefully, this situation will change in the near future as the small wind industry adopts standards for performance and wind speed rating for small wind turbines.

Consider three wind turbines with identical rated power of 1,000 watts or 1 kilowatt: (1) Bergey's XL.1, (2) Southwest Windpower's Whisper 200, and (3) Eveready's Kestrel 1000. At first blush, you might assume that all three wind turbines are identical. After all, they're one kilowatt turbines. Upon closer examination, you'll find that Southwest Windpower's Whisper 200 achieves its 1,000-watt rated output at 26 miles per hour. Bergey's XL.1 produces its 1,000-watt rated power at 24.6 miles per hour, and Eveready's Kestrel cranks out 1,000 watts at 23.5 miles per hour. So which one would be best?

Although it might seem that the wind turbine that produces 1,000 watts at the lowest rate wind speed would be the best buy, that's not necessarily the case. There's much more to consider, including swept area, top-of-tower weight, AEO, governing mechanism, and so on.

In this example, the swept area of the Bergey BWC XL.1 is 53 square feet. The Whisper 200's swept area is a bit larger, 63.5 square feet. Topping the group, however, is Kestrel. Its swept area is 79 square feet. Which one will produce the most electricity for you?

As you learned earlier, in most cases, the wind turbine with the largest swept area produces the most electricity, but don't make that assumption without looking at the annual energy output. In this example, the Kestrel does indeed out-produce its competitors, as predicted. At an average wind speed of 8 miles per hour, the Kestrel 1000 will produce about 900 kilowatt-hours of electricity annually. The Whisper 200 will produce about 720 kilowatt-hours annually and the Bergey will produce approximately 660. (These numbers are based on the manufacturers' estimates.)

Rated power is only one point on the power curve and is the output at a wind speed that is much higher than those typically encountered at most sites. Thus, the rated wind speed only represents a small fraction of the wind resource at a site. We're more interested in output in the typical range of wind speeds at our site, which is well below the rated power. Our advice on rated power: forget it.

Final Factors

When shopping for a wind turbine, be sure to check into the company's customer service record — how well it supports its dealers and customers. Some companies have notoriously poor customer service. Others like Bergey Windpower and Abundant Renewable Energy have stellar records.

Another factor to consider is how long the company has been in business. As a general rule, it's a good idea to do business with companies that have been around for a long while. (Bergey's been in business since the early 1980s.)

Yet another criterion to take into consideration is how long a wind turbine has been on the market. "I'll install version 2.0 for a customer, maybe, but not version 1.0," notes

Aaron Godwin of The Renaissance Group. Wind turbines that have been on the market for a while have been tried and tested and usually improved upon.

We recommend against buying a turbine in its first year or two on the market. Many people who made this mistake have been extremely disappointed. The new wind turbines had often not been fully tested and debugged and the unsuspecting buyer ended up being beta testers.

When you consider different wind turbines, it's also important to check on the availability of parts. Parts shipped from foreign locations, such as China, Europe or South Africa may take months to arrive, if they're available at all. Meanwhile, your $30,000 wind energy system sits idle. When shopping, check to see if the importer of the wind turbine you are interested in stocks spare parts or spare turbines to supply any parts quickly.

Speaking of parts, when shopping for a turbine, it's a good idea to consider the number of moving parts and wear points in a turbine. Some designs, like the Jacobs turbines manufactured by Wind Turbine Industries in Minnesota, contain a lot of moving parts resulting in numerous wear points — up to 300. These machines may require a lot of tinkering at 100 feet. They're not for the acrophobic.

While you are at it, be sure that the company you buy from offers technical advice should you have trouble installing the turbine or if the turbine breaks down. And do their technical support staff speak your native tongue? Although many French and Spanish turbines are tried and tested by companies that have been in the business for a long time, if you live in North America you may still have language and distance barriers to overcome. (Some foreign turbines are imported and may be reliably supported by importers.)

While you are at it, check into the warranty. Warranties typically run five years. The longer the better. Also be sure to determine what they cover. Is it materials and workmanship or parts only? Does it include shipping? Unfortunately, most warranties do not cover the cost of labor.

When buying a wind turbine for a grid-connected system, be sure that the inverter is compliant with Underwriter's Laboratory requirements (UL 1741) or its equivalent. This ensures safe interconnection with the grid. Note that UL does not test or certify wind turbines

What Voltage?

As you shop for wind turbines for wind systems with battery backup, you'll find that they come in various voltages — typically, 12, 24 and 48 volts. Which voltage is best for your application?

As a rule, we recommend higher-voltage wind systems for modern homes. (Lower voltages may be appropriate for small cabins and boats.) Higher voltage systems are more efficient, especially if long wire runs are required. They are more efficient because higher voltage results in lower current, and lower current results in lower losses in the wire. (Wire losses are a function of the amount of current in a wire. The higher the current, the higher the losses. The lower the current, the lower the losses.) Current can be decreased by increasing the voltage.

Why does an increase in voltage decrease the current through a wire?

This can be explained by referring to the equation: watts = amps x volts (sometimes called the power formula). If 12 amps of electricity are flowing at 12 volts, the wattage is 144. If the voltage is increased to 24, and the wattage remains the same, the amperage drops to 6. If the voltage is increased to 48, and the wattage remains the same, the amperage drops to 3. As a result, a 48-volt system can use much smaller wire than a 24-volt system. A 24-volt system can use much smaller wire than a 12-volt system. For a mathematical explanation of this phenomenon, see the sidebar on this page.

Current and Wire Losses

Wire losses are proportional to the resistance of the wire and the square of the current. If you double the voltage, the current is cut in half and wire losses are reduced by a factor of four. For long wire runs, higher voltage also means that you can avoid using expensive heavy-gauge wire.

at this time. While lack of certification is not an indication of a lack of safety, it may cause problems for you when trying to get approval to connect your system to the utility. We show you how to deal with these in Chapter 10.

We recommend buying from a local dealer/installer, too. That way, you'll have local support, even if you opt to do the installation yourself. Be sure to ask for references and interview them.

If you buy from an online wholesaler, be aware that most of these companies do not offer technical support, installation advice or assistance. Nor do they offer replacement parts or repair services. If you buy cheap, you are on your own when you need help.

We believe that it is a good idea to stick with name brand wind turbines and avoid newcomers and foreign imports, especially Chinese-made machines, at least at this time.

Buying a used or reconditioned wind turbine may be an economical option, and will typically cost about 50 percent to 60 percent less than buying a new turbine — potentially saving you a huge amount of money. The problem with used wind turbines is that you don't always know what you're getting and, if you did, you wouldn't be happy. "Used machines from early California wind farms have been beat to death," notes Paul Gipe, "eBay is not a good place to buy a used turbine that works," notes Aaron Godwin. If the turbine is not reconditioned, it may cost a fortune in time and parts to rebuild it. Only the foolish install a non-rebuilt wind turbine, and they usually only make that mistake once.

If you buy a remanufactured wind turbine, like a Jacobs 31-20 or the larger Vestas from individuals or companies that refurbish these workhorses, "be sure that the machine has been fully reconditioned, not just painted and had new blades installed," advises Godwin. Also, when buying one of the larger reconditioned wind machines (25-foot and longer blades), be sure that the controller or inverter that comes with the unit meets current standards for utility connection. You may also want to obtain an extended warranty. Also be sure you can find a tower for the turbine *before* you purchase it. Suitable tall towers for the larger small-scale wind machines are hard to find.

Although a few good deals on used turbines are available from time to time, we recommend extreme caution when considering this route. Shop very carefully. When it comes to issues of maintenance, performance, and the reliability of a used machine, you have only the word of the seller to go on.

Building Your Own Wind Machine

If you would like to tap the power of the wind, but don't have the cash to buy a system or you love to tinker and want the challenge of making your own turbine, you may want to consider building your own wind turbine (Figure 5.13). Building your own wind generator provides invaluable experience — and helps you appreciate all of the hard work that has gone into residential wind turbine design over the years. And, Ian notes, "it really helps you understand how a wind turbine works!"

Fig. 5.13:
*This amazingly
quiet wind turbine
designed and built
by the folks from
Otherpower.com is
remarkably reliable
and efficient.
Otherpower.com
sells kits that require
some assembly for
independent and
budget-constrained
buyers.*

Directions to build a wind machine are available online. You can locate plans and advice on building your own by searching for "building your own wind machine."

In your search, you will very likely encounter an article by Steve Hicks, entitled "Building Your Own Wind Generator." It's available through the American Wind Energy Association's web site and is well worth reading. There are even a couple of books on the subject: *Windpower Workshop* by Hugh Piggott and a more recent title, *Homebrew Wind Power: A Hands-On Guide to Harnessing the Wind* by Otherpower.com's Dan Bartmann and Dan

Fink, an extremely well-written and thorough treatise on the subject.

You can also go directly to a couple of useful websites: Hugh Piggott's scoraigwind.com or Otherpower.com. At these sites, you'll find a listing of other valuable websites and resources. Homebrew wind turbine master Hugh Piggott's web page sells step-by-step plans for building axial-flux alternator turbines.[4] Otherpower.com offers free plans for one turbine, as well as parts and kits — that is, almost completely assembled homebrew wind turbines. Their turbines were inspired by Hugh Piggott. Included in the kit is the alternator, blades, tail vane, all necessary hardware to assemble the machine, and a user's manual. "All that remains for you to do," write the owners of the company, "is to apply finish to the wooden parts (blades, blade hubs and tail vane) and assemble the turbine. You will also need to provide your own tower, three-phase rectifier, diversion regulator and all other parts of the system." This machine can charge batteries in 12-, 24- or 48-volt systems. (You need to stipulate system voltage when you buy the turbine kit.)

You can also buy parts and/or complete kits, through several online suppliers, including windstuffnow.com and mikeswindmillshop.com. We recommend that you sign up for a workshop, for example, through the Midwest Renewable Energy Association, taught by the folks at Otherpower.com.

Rather than provide a full description of this subject, which would take a chapter or two, maybe even an entire book, we recommend you look at these resources. We'll simply make

a few salient points and let you decide if you want to build your own machine.

Building a wind generator is not that difficult, if you are mechanically inclined and persistent. You will need a shop and some common hand and power tools, and you'll need to be able to think on your feet. As a rule, wind machines with 12-foot or larger blades are beyond the abilities of most home builders, requiring very strong construction skills. (However, the folks at Otherpower.com have recently introduced a kit for a homebrew wind turbine with 12-foot blades.) Smaller turbines, with 8–11-foot blades, are easier to build.

"The hardest part of designing a small windmill for electricity production is to find a suitable generator," writes Piggott in *Windpower Workshop*. "For best performance you'll need a reliable, low-speed generator that's pretty efficient in light winds," he adds. Piggott strongly recommends permanent magnet alternators. "They are," he says, "by far the most popular choice on successful small windmill designs." Their alternators do not require brushes or slip rings. The only parts that can wear out are the bearings.

Unfortunately, permanent magnetic alternators are difficult to find. Piggott lists some sources, among them motorcycle alternators and welders. Unfortunately, each one has a significant downside. Motorcycle alternators, for instance, are designed for high rpm performance and are not easily adaptable to direct drive wind turbines.

Another option, says Piggott, are brushless DC motors. Brushless DC motors are like permanent magnet alternators in many ways and are an almost perfect choice, too. They're currently used in some new washing machines, machine tools, and medical equipment.

Car alternators are popular among do-it-yourselfers. They are widely available and are fairly inexpensive, even if purchased new. They are also designed to charge batteries. Unfortunately, car alternators require high rpms and are not always very efficient. They require a lot of electricity to power the electromagnets that create the magnetic field. Moreover, if you go this route, you'll need to rewind the stator coils (these are the coils that produce electricity), at least doubling the number of windings. This enables the alternator to produce electricity at lower speeds. Even then, the power output curve of a car alternator is poorly matched to the power curve of a wind turbine rotor, resulting in poor performance in low winds and out-of-control rpm at higher wind speeds. "Most people who have tried adapting car alternators to wind turbines have given up in frustration. Their underperforming wind turbines are strewn across the countryside after the first good storm," notes Mick.

Building a generator is challenging and vital to the success of a homemade wind turbine. But it's not all you'll have to do. You'll have to design and build blades and install controls.

Blades and rotors are particularly tricky. Because they are subject to considerable stress, they need to be well constructed so they don't break apart. "For minimum stress, the blade should be light at the tip and strong near the root," recommends Piggott.

Most homemade wind turbines use wooden blades. Wood is light, strong and readily available. It is also easy to fashion into a workable blade and resistant to stress. Metal, on the other hand, is prone to stress and steel is heavy. Fiberglass is great, but most homeowners don't have the equipment or expertise needed to mold fiberglass blades.

When designing blades, you'll need to select the rotor diameter very carefully. The diameter should match the operating speed of the alternator. Piggott's book and website provide excellent advice on the subject, along with advice on carving, painting and balancing blades.

A rotor and a generator do not a wind energy system make. You'll also need to devise a frame and tail vane and install necessary controls. Overspeed controls and shut down controls, discussed in this chapter, should be incorporated, if at all possible. Otherwise your wind turbine could be destroyed by the first storm that passes through your area. A charge controller (discussed in Chapter 3) should also be installed in battery-based systems to prevent batteries from being overcharged. You'll also have to install a dump load, a resistor to dump excess electricity when the batteries are full.

As with any wind turbine, you'll also need to install the turbine on a suitable tower. Tower designs can also be found online. Be careful, however; homemade towers often leave much to be desired. Towers are far more complicated to design than wind turbines. Your best and safest bet is to buy a used tower, if you want to save money.

Building your own turbine is fun and challenging, and can save a lot of money. Moreover, there's lots of good advice on the subject. And although we'd never discourage anyone from giving it a try, we'd be remiss if we did not point out that homemade wind turbines may not produce as well as wind generators sold by any of the world's best manufacturers, such as Bergey, Proven and Abundant Renewable Energy.

The best manufacturers design, engineer and build their small wind machines with performance and reliability in mind. Many companies have been relentless in their pursuit of quality and continue to improve their products, making the most reliable and efficient wind generator humanly possible. Homemade wind machines generally can't match these highly refined, finely tuned machines, but they're a lot of fun to build and may even provide some of your electrical needs! The only exceptions that we have seen are the designs available from Hugh Piggott and Otherpower.com. Dan recently purchased a 12-foot turbine made by students at a Homebrew workshop taught by Otherpower.com. Early tests revealed that the turbine runs extremely quietly and produces energy on par with manufactured small wind turbines. Even though they may not be as technologically sophisticated as modern wind turbines, well-made homebuilt wind turbines — like the designs from Otherpower.com — can last for many years if properly maintained, and, in Ian's experience, sometimes survive and perform better than light- or medium-duty manufactured turbines. And if you build your own, you'll really understand your wind generator!

Conclusion

By now, you realize that the simple wind turbine you've been eyeing on a neighbor's property is a rather complicated and fine-tuned machine. Moreover, you may have discovered that buying a wind turbine won't be as easy as you may have thought. There are a lot of factors to take into account. Price is only one of them. Be sure to choose a machine that meets your needs. If you're installing a grid-connected system, choose a turbine that works with such systems.

If you're installing a battery-based system, look for a battery-charging turbine.

Also, take into consideration key factors like the swept area of the turbine and the annual energy output. If you want a good machine, one that will last for many years, you'll need to look for one of the meatier models. It will cost more initially, but will outlast its cheaper cousins, costing much less in the long run.

CHAPTER 6

TOWERS AND TOWER INSTALLATION

Many people considering wind energy focus their attention on the turbine and, therefore, pay little attention to another essential component of a successful wind system, the tower. Focusing attention on the wind turbine is quite natural. After all, it's the collector, that part of the system that captures and converts the wind's energy into electricity. Sitting high atop their towers, spinning in brisk energy-rich winds, turbines are the superstars of wind systems.

As any owner of a successful wind system will tell you, however, it is not just what's on top of the tower that counts. The tower itself plays a huge part in creating a successful wind energy system. The tower not only raises a wind turbine to a height at which it can harvest lots of energy from the wind, it also withstands nature's fury. In addition, the tower can be quite costly. In large systems, the cost of a tower may account for nearly a quarter of the

system cost. For smaller wind turbines, an appropriately sized tower plus installation may cost two to three, sometimes five times, more than the turbine.

Tower installation involves a significant investment in time, money, and energy. Most students at wind energy workshops assume that they'll be spending a lot of time working on the wind generator. They're often surprised to find that's not the case. "Most of the installation time of a six-day wind workshop is spent assembling and raising the tower," Ian notes in his article "Wind Generator Tower Basics" in *Home Power* magazine (Issue 105).

During a wind energy system installation, it can take one to four days to excavate and build a foundation for a tower and another two to five days to assemble it. Installing inverters, controllers, and system wiring, especially for battery-based systems, accounts for another significant portion of the installation time. Assembling the turbine, attaching the turbine to the tower, and making the electrical connections may only require two to four hours.

Given the importance of the tower, its cost, and other considerations, it is vital that you understand a bit about towers *before* you shop for a wind turbine. In the pages that follow, we'll explore three tower options. We'll briefly describe tower assembly and installation. We'll discuss proper siting to help you understand where a tower and wind turbine should go on

A Primer on Wind Generator Tower Physics

Suppose you are getting ready to paint the kitchen in your home. You're set to go, but discover that you have forgotten the screwdriver needed to pry open the paint can lid. It's in your toolbox in the garage. It's ten degrees below zero outside and you're not feeling like braving the cold. What to do?

You could bundle up and trudge out to the garage to retrieve it or you could fish around in your pocket for a quarter to open the paint can. Since the quarter in your pocket involves a lot less work than retrieving a screw driver from your workbench and you are raring to go, you opt for the quarter. What you find, however, is that you expend at least as much effort prying open the lid of the paint can with a quarter as it would have taken to retrieve a screwdriver from the garage. Why?

To open the paint can, you need a device that physicists call a lever. Both a quarter and a screw driver can serve as levers to pry open the lid. Although both "tools" will work, much more force is required to open the paint can with a quarter than a screwdriver. Why?

In a lever, a force is applied to one end and the lever transfers that force to some object at its other end. How the force applied to the lever compares to the force transmitted to the object depends on the position of the pivot point. If the pivot is in the center of the lever, the forces will be equal. But if the pivot is closer to the object (load end), the force at that end will be greater than the applied force.

The equation that describes this is: $f_1 \times d_1 = f_2 \times d_2$, where f_1 is the applied force and d_1 is the distance from the applied force to the pivot, f_2 is the force at the load end and d_2 is the distance from the pivot to the load end. Rearranging the equation to solve for the force at the load end gives: $f_2 = f_1 \times d_1/d_2$. Since the distance from the edge of the paint can (the pivot point) to the lip of the lid (d_2) is the same for both the quarter and the screwdriver, but the screwdriver has a much longer lever arm (d_1) than the quarter, the screwdriver will deliver much more force to the lid of the can.

So, what does all of this have to do with a tower for a wind turbine? It's quite simple. It turns out that the wind views the tower as a lever, and the wind applies a force to one end of the lever, the top of the tower. The downwind pier of the foundation, shown in Figure 6.1, is the pivot point, and the upwind pier of the foundation is the other end ☞

your property. We'll discuss the economic benefits of installing a tall tower vs. a short tower. We'll also examine other key economic issues, like whether it makes sense to put a larger turbine on a short tower and whether two smaller turbines make more sense than one turbine on

a tall tower. We'll end with a description of safety concerns and give you some advice on buying a wind generator tower.

Before we begin our exploration of the towers, it is useful to discuss the forces that understand why it is important to buy — or, if

of the lever. In a tower where the height is ten times the distance between the downwind and upwind piers, a wind force applied to the top of the tower will be multiplied by ten. This force acts to lift the upwind foundation out of the ground.

The amount of force the wind can apply is directly related to three factors. The first is the wind speed. The greater the wind speed, the more force it applies. The second and third factors are the swept area of the rotor and the frontal area of the tower. The greater the area of the rotor and tower, the more force the wind applies to lifting the foundation.

"As swept area, frontal area and wind speed (force) increase and tower height (distance) increases, so does the overturning force exerted by the wind on the tower and its concrete footings," notes Mick.

For a successful wind system, then, the tower needs to be strong enough and anchored to the ground well enough to thwart the wind's potentially destructive force. Never underestimate the power of the wind. ■

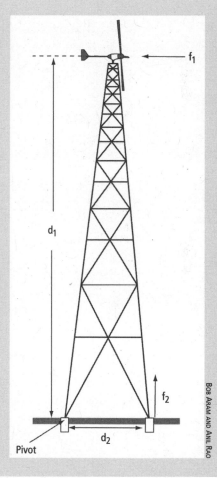

Fig. 6.1: *Forces acting on a tower.*

BOB ARAM AND ANIL RAO

Is Wind Right for You?

Periodic inspection and maintenance and occasional repair of wind turbines are essential to the long-term success of a wind energy system. The towers on which they stand present a formidable barrier to these activities. Many wind system owners fail to perform these tasks because they don't want to lower or climb their towers once a year. So, if you are a "put it up and forget about it" kind of person and can't afford to hire someone to perform an annual inspection and maintenance, we recommend that you consider installing a PV system instead of a wind system. PV systems are as close to maintenance-free technology as you can get (provided there are no batteries in the system).

If you install a wind system, you will either need to climb the tower or lower it to the ground at least once a year — some experts recommend twice a year — to inspect the turbine, wires, connections, and perform maintenance, as required. (We'll describe this process in more detail in Chapter 9.)

Table 6.1 Tower Types

1) Freestanding
   a) Lattice
   b) Monopole

2) Fixed Guyed
   a) Lattice
   b) Tubular (typically homebuilt)

3) Tilt-up
   a) Guyed tubular
   b) Guyed lattice

you're a do-it-yourselfer build — a high-quality tower. It will also help you understand the importance of properly designed foundations and anchors. In order to understand this, we present an example Mick uses in his wind energy workshops in the accompanying box. It tends to pry a wind tower from its foundations. This topic is covered in the box on page 146.

Tower Options

Towers for small wind machines come in three basic varieties: (1) freestanding, (2) fixed guyed, and (3) tilt-up towers. Each type has some variations, listed in Table 6.1. Figure 6.2 shows what each looks like. Let's take a look at each one.

Freestanding Towers

As their name implies, freestanding wind generator towers are self-supporting structures. They stand on their own, like flag poles or street lights, or the Eiffel Tower in Paris. Freestanding towers for small wind turbines are made of steel and are firmly anchored to the ground by well-reinforced concrete foundations, discussed shortly. They do not require wires or steel cables (known as guy cables or guy wires, not "guide wires") to hold them up. The combination of heavy-duty steel tower construction and a secure anchorage ensures that the tower can withstand powerful winds that, if they had their way, would pry the foundation loose and topple the tower. They also ensure that the tower can support the turbine.

The most common type of freestanding tower is a lattice or truss tower, like the one shown in Figure 6.3a. The Eiffel Tower in Paris is a good example of a lattice structure.

Lattice towers are made from tubular steel or angle iron with horizontal and diagonal cross-bracing. The cross-bracing is bolted to

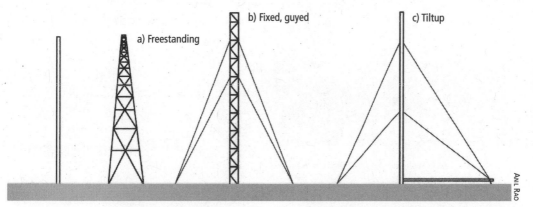

a) Freestanding b) Fixed, guyed c) Tiltup

Fig. 6.2:
*Wind Tower Options.
(a) Freestanding
towers, (b) fixed,
guyed towers, and
(c) tilt-up towers.
Freestanding
towers can be
both lattice and
monopoles (shown
here). Fixed, guyed
towers are typically
lattice towers.
Tilt-up towers
can be either
lattice or tubular.*

a

b

Fig. 6.3:
*Freestanding lattice
towers are made of
(a) heavy-duty
angle iron, as in
this tower, or (b)
tubular steel.
Horizontal and ver-
tical bracing made
of steel angle iron
that attaches to the
tubular steel legs.*

PROVEN ENERGY

Fig. 6.4: *Monopole towers are sturdy, well-anchored by a solid foundation, but extremely costly for reasons explained in the text.*

MICK SAGRILLO

Fig. 6.5: *This massive, deep foundation was built to support a 120-foot freestanding lattice tower. Each leg of the tower will attach to a steel leg embedded in each of the vertical piers. The piers and base of the foundation are made of concrete reinforced with rebar.*

the vertical steel legs. Ladders are often incorporated into the design so the towers can be climbed for inspection, maintenance and repair. In some lattice towers, step bolts are installed on one vertical leg for climbing.

Larger freestanding lattice towers are sometimes fitted with a small platform near the top. The platform provides a secure place to work. Workers should wear safety harnesses and should be secured to the tower via lanyards (more on this shortly).

Another, albeit considerably more expensive option for freestanding towers is the monopole, a single, sturdy pole (Figure 6.4). These sleek towers are made from round tubular steel. Rungs or foot pegs are included on the tower for climbing.

Lattice and monopole towers are secured to massive foundations made of concrete and reinforced with steel (rebar), as shown in Figure 6.5. In a lattice tower, each leg is anchored to its own foundation pier. In some designs, the piers that support the legs are connected to a block of concrete at their base. In a monopole tower, there's one central foundation for support. The taller the tower, the heftier the foundation.

Assembling and Installing Freestanding Towers

Freestanding lattice towers are typically assembled on the ground in sections (Figure 6-6b). The Jacobs 31-20 tower the Midwest Renewable Energy Association installed in the fall of 2007, for instance, was assembled 20 feet at a time. The legs and bracing were bolted together on the ground.

a

After a lattice tower is assembled, the turbine is often attached. The tower and turbine are lifted with a crane. The tower is then bolted to steel anchors embedded in the concrete foundation. In some instances, the tower is erected without the turbine. The turbine is then hoisted onto the top of the tower and attached.

To facilitate tower construction, some lattice towers are hinged at the base figure 6.6a and c. This allows workers to assemble the tower on the ground, then tilt it up into position using a crane.

Cranes must have a lifting capacity sufficient to handle the tower or the combined weight of the turbine and tower. They must also have sufficient height and reach (Figure 6.7). We used an 80-ton crane to lift the 120-

c

Fig. 6.6:
Hinged Lattice Tower.
(a) This lattice tower is hinged at the base, allowing workers to (b) assemble the tower on the ground. Once the tower is assembled and the turbine is attached, the tower is tilted into position using a crane. (c) Another view of the heavy duty hinge that makes all this possible.

foot tower and Jacobs 31-20 turbine, weighing in excess of 5 tons, at the MREA headquarters, but needed a heavy-duty forklift to assist.

A reasonably level area is also needed to assemble a freestanding tower and to lift it

Fig. 6.7: *Crane Lifting Tower and Turbine. This 80-ton crane lifts a massive turbine and tower into place. The hinges at the base of the tower allow the crane to tilt the tower into position.*

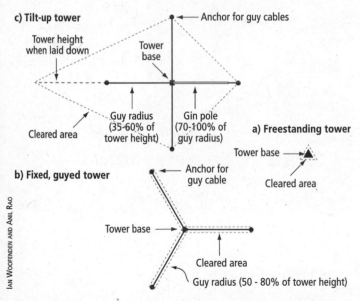

c) Tilt-up tower

Tower height when laid down

Tower base

Anchor for guy cables

Cleared area

Guy radius (35-60% of tower height)

Gin pole (70-100% of guy radius)

a) Freestanding tower

Tower base

Cleared area

b) Fixed, guyed tower

Anchor for guy cable

Tower base

Cleared area

Guy radius (50 - 80% of tower height)

IAN WOOFENDEN AND ANIL RAO

Fig. 6.8: *Footprint of Wind Generator Towers. (a) Tilt-up, (b) fixed, guyed, and c) freestanding.*

with the crane. Care must be taken when initially lifting the tower to prevent the blades of the turbine from breaking. This is accomplished by rotating the turbine so the hub faces the ground. (That way, it won't swing around and snap the blades as the tower is lifted.) Two workers hold the blades in place as the turbine is initially raised.

Although freestanding lattice towers are typically assembled on the ground and lifted with a crane, it is possible on inaccessible sites to construct towers vertically one section at a time using a vertical gin pole. This technique is time-consuming and requires extreme caution.

Like lattice towers, monopole towers come in sections. They are fitted together on the ground. When completed, the tower is hoisted into place with a crane. The tower is anchored to the ground by a deep steel-reinforced concrete foundation.

Pros and Cons of Freestanding Towers

Freestanding towers offer advantages over towers supported by guy cables. One of the most important is that they require much less space (Figure 6.8). Their smaller footprint makes a freestanding wind generator tower ideal for locations with extensive tree cover. If space is at a premium, give the freestanding tower serious consideration.

Freestanding towers are more aesthetically appealing to many people than guyed towers. A freestanding tower is also one of the safest towers to install, if you know what you are doing, of course. Almost all the work can be done on the ground, and a single crane lift can

Table 6.2 Estimate of Tower Costs for a 120-foot Tower*			
Tower Type	**Tower Cost**	**Installation Cost**	**Total Cost**
Freestanding monopole	$25,000	$25,000	$50,000
Freestanding lattice	$15,000	$12,000	$25,000
Guyed tilt-up	$10,000	$4,000	$11,000
Guyed lattice	$8,000	$2,000	$9,000

*Based on 2007 cost data for a typical 10 KW turbine from Mick Sagrillo's Wind Site Assessor Course.
These prices include not only the cost of the tower, but also shipping, sales tax, excavation, rebar, concrete, labor to form the foundation, pour the foundation, and strip the forms, and finally backfilling.

erect the tower, turbine, wiring, etc. However, one worker needs to climb the tower to release the crane attachment, after the tower is bolted to the foundation.

Because of the large amount of steel, the need for a solid (deep) concrete foundation, and labor to install them, freestanding monopole towers are typically the most expensive of all options (Table 6.2). A 120-foot freestanding monopole, in 2007, cost about $25,000 plus another $25,000 to install.[1] Freestanding lattice towers cost less, about $15,000 plus $10,000 to install. Both are out of the reach of most homeowners. Why are freestanding towers so much more expensive than guyed towers?

Freestanding towers require a considerable amount of steel and concrete to withstand the wind's overturning forces. This results in a more expensive tower. In contrast, all guyed towers use trigonometry to their favor. In a guyed tower, the guy cables shift the loads exerted upon the tower base away from the tower. As a result, the amount of steel required in guyed towers is much lower than in freestanding towers.

Because they require so much concrete and steel and because these materials require huge amounts of energy to produce, freestanding towers also have a much higher embodied energy than other options. Embodied energy is the energy that it takes to make a product — from the extraction of the raw materials to the completion of the finished product, including shipping to retail outlets where it is sold. If your primary motivation is to decrease your environmental footprint by using renewable energy, think twice about how much energy it takes to manufacture and install various tower options.

Another potential downside of freestanding towers is that they require periodic ascent to perform routine inspection, maintenance and repair. To prevent catastrophic falls, a safety harness or safety work belt must be worn (Figure 6.9). Towers should be equipped with a safety

DAN CHIRAS

Fig. 6.9: *Safety Harness. Mick demonstrates proper use of safety harness in one of his workshops.*

Fig. 6.10: *Safety Cable and Lad-Saf. (a) Worker prepares to climb tower. Note safety harness and Lad-Saf attached to safety cable. This prevents the worker from falling. (b) Close up of Lad-Saf and safety cable.*

DAN CHIRAS

a

DAN CHIRAS

b

cable that runs the length of the tower along the climbing rungs or ladder (Figure 6.10). Workers attach their safety harness to the cable when climbing by a device known as an anti-fall mechanism, such as a Lad-Saf, one commercially available product. It allows the worker to ascend and descend the tower, but locks if the worker loses his or her footing and falls.

As shown in Figure 6.9, safety harnesses comes equipped with several D-rings. They're used to secure lanyards that are attached to the tower to prevent falls when working on a tower. A "positioning" or "restraint" lanyard holds a worker in place to allow him or her to work hands-free. A "shock absorbing" lanyard arrests a fall, slowing the worker gradually to prevent a harmful jerk.

Even though good safety equipment prevents injury, many people don't like the idea of climbing a tower. If you are not willing or able to climb a tower, you must be willing to hire someone to do it. If not, consider installing a tilt-up tower or PVs.

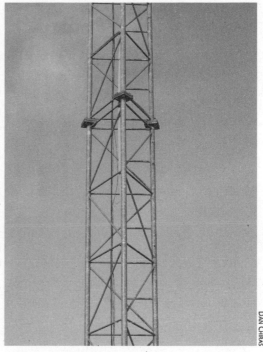

a   b

Fig. 6.11:
*Fixed Guyed
Lattice Tower.
(a) This lattice
tower is anchored
by guy cables. This
is one of the most
popular and least
expensive tower
options. (b) Close
up showing details.*

Fixed Guyed Towers

The second major type of tower is the fixed guyed tower (Figure 6.11a). They are often used for radio and television antennas. Most lattice towers consist of triangular lattice pieces that come in 10- or 20-foot pieces. The legs of fixed guyed towers are made of steel tube or pipe, or sometimes solid steel rods. The three legs of the lattice tower are usually 18 inches apart and are secured by horizontal and diagonal steel cross braces (Figure 6.11b). Unlike freestanding lattice towers, the legs of a guyed lattice tower do not taper from base to top. To inspect and maintain the tower and turbine, you'll need to climb the tower.

Guyed towers are bolted onto a concrete foundation (base) and are supported by guy cables (Figure 6.11a). Guy cables are made of high-strength stranded-steel cable or aircraft cable. They extend from attachments on the tower to anchors embedded in the ground. Guy cables are strung out in three directions and are anchored 120 degrees apart. The guy radius, that is, the distance from the base of the tower to the anchors, is usually about 75 percent of the tower height, but can range from 50 to 80 percent, depending on the construction of the tower. For a 120-foot tower, then, the anchors would be 120 degrees apart and 60 to 100 feet from the base of the tower.

Fixed guyed towers are also made from pipe or tubular steel which comes in 20-foot sections. Like guyed lattice towers, tubular towers are supported by steel guy cables.

Assembling and Installing Fixed Guyed Towers

Guyed towers are usually assembled on the ground. Lattice towers are bolted together. For tubular towers, one section of pipe slips into or bolts onto the end of the next section.

After the tower is assembled, the wind turbine, electrical wire and guy cables are attached. The tower and turbine are then erected by a crane. In some cases it may be necessary to lift a lattice tower or tower made from steel tubing in sections, depending on its height and strength. That's because some towers are rather flexible and may not be strong enough to permit the tower to be lifted all at once. This is typically encountered when the tower is 80-feet or longer. In such cases, the wind turbine is lifted onto the tower after the last section is in place.

Fixed guyed towers can also be assembled vertically, one section at a time, using a vertical gin pole. This type of gin pole is an inexpensive and temporary vertical crane. It is bolted onto the tower and is used to raise sections of the tower one at a time. After a section is in place, the gin pole is moved up so the next section can be then installed, and so on. Vertical gin pole assembly is time-consuming and tedious and can be a bit dangerous. Those who've tried, often recommend against it. If no crane is available or the crane cannot access the site, however, a gin pole may be your only option.

Fixed guyed towers are supported by concrete foundations, though they are generally not bolted to them. Guy cables are attached to the tower prior to erection. Once the tower is upright and plumb, workers tension the cables. If the turbine was not previously attached, it is then lifted by the crane and fastened to the top of the tower.

Pros and Cons of Fixed Guyed Towers

Fixed guyed towers cost much less than freestanding towers because they require much less steel and their foundations require a lot less concrete than those of freestanding towers. Lattice towers used by installers are also mass produced for the telecommunications industry, making them less expensive and widely available. Fixed guyed towers require more space than freestanding towers but much less space than tilt-up towers, discussed next.

Fixed guyed towers must be climbed for routine inspection, maintenance and repair, like freestanding towers. And, of course, there's the issue of aesthetics. Some people view the guy cables as a bit of an eyesore, although we should point out that guy wires disappear into the background from most vantage points, except up close. Guy wires may also present a hazard to birds. "Birds can fly into the wires by accident because they are small diameter and difficult to see," Jim Green points out, although such occurrences are rare.

Tilt-Up Towers

The third type of tower option is a guyed tilt-up tower. As their name implies, guyed tilt-up towers are supported by guy cables. Unlike freestanding and fixed guyed towers, guyed tilt-up towers can be raised and lowered to allow an owner to perform inspection, maintenance and repair.

Guyed tilt-up towers come in many heights — up to around 130 feet (40 meters). They are most often made from steel pipe, although guyed tilt-up lattice towers are available.

Guyed tilt-up towers require four sets of guy cables at each level. Cables are located 90 degrees apart. As in the fixed guyed tower, guy cables of guyed tilt-up towers provide support. The fourth set of cables, however, enables workers to safely raise and lower the tower. Without the additional set of cables, the tower would topple as it is raised or lowered.

As illustrated in Figure 6.12, a tilt-up tower is raised and lowered with the aid of a gin pole. Unlike the vertical gin pole used to erect a tower, the gin pole in guyed tilt-up towers is permanently attached to the mast (the pole or lattice tower). This steel pipe (or a section of lattice tower) is permanently attached to the base of the tower at a right angle (a 90° angle) to the mast. It serves as a lever arm that allows the tower to be tilted up and down for inspection, maintenance, and repair.

Tilting a tower up and down is made possible by a hinge anchored to the concrete base (Figure 6.13). This hinge forms a joint between the foundation and the base of the tower that operates much like an elbow joint. When the tower is down — that is, lying on the ground and ready to be raised — the gin pole sticks straight up. When the tower is vertical, the gin

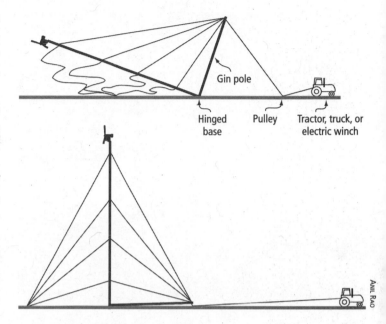

Fig. 6.12: *Guyed tilt-up towers are raised and lowered using a truck, tractor, electric winch or grip hoist.*

Fig. 6.13: *The hinge at the base of this tilt-up tower allows it to be tilted up and down to maintain and service the wind turbine.*

pole lies near and parallel to the ground. As illustrated in Figure 6.12, a steel cable connects the free end of the gin pole to a lifting device such as a tractor, which is used to raise and lower the tower.

As in fixed guyed towers, guy cables hold a tilt-up tower upright and resist the force of the wind. Guy cables typically connect to each section of tower. The taller the tower, the more sets you'll need.

As in fixed guyed towers, guy cables attach to anchors. The guy radius is 35 to 80 percent of tower height. For a 120-foot tilt-up tower, then, the anchors should be located 42 to 92 feet from the base of the tower. The fourth guy cable is permanently attached to the end of the gin pole, as shown in Figure 6.12. The free end of the gin pole is connected to the fourth anchor as shown in Figure 6.14.

Assembling and Raising a Tilt-Up Guyed Tower

The steel pipe or tubing typically used for guyed tilt-up towers comes in 20-foot lengths. The individual lengths are secured by bolts or joined by slip-fit couplings on the ground. While the tower is on the ground, guy cables are attached to the tower and the concrete anchors.

Once assembled, the tower is tilted into position (Figure 6.12). This is accomplished with the assistance of a tractor, a pickup truck, a heavy-duty electric winch, or a manually operated grip hoist.[2] The lifting device is attached by a cable to the end of the gin pole. Although they're slow compared to other options, grip hoists are safe and affordable (Figure 6.15). "They provide a level of control that no other lifting method offers," says Ian. "Faster methods can lead to disaster before workers notice a problem. Slow and steady sometimes is best."

When installing a tall tower for the first time, some installers raise one or two sections of the tower at a time. After one section is raised, the tower is plumbed and the guy cables are properly tensioned. This section is then lowered so that an additional segment can be added. It is then raised, plumbed, and cables tensioned. This process continues until the entire tower is assembled, plumbed and properly tensioned.

Experienced installers also recommend lifting (and plumbing) the entire tower *before* attaching the wind turbine just to be sure that everything is working. As Ian notes, "Tilt-up towers *will* be 'noodley' and out of plumb when first raised. You don't want the added weight of a wind turbine to deal with. And you don't want to risk damaging a wind turbine should the tower collapse because it was not plumbed or the cables were not tensioned properly, or something else went awry." Lifting a tower initially without the turbine is also important because it helps train your workers. That way, everyone knows what they are doing by the time the turbine and tower are lifted. Also be sure all workers and onlookers, including family pets, are out of harm's way.

Pros and Cons of a Guyed Tilt-Up Tower

"My advice: if you have space for a tilt-up tower, use one!" advises Ian in "Wind Generator Tower Basics" in *Home Power* magazine. "You will never have to climb your tower (in fact, you won't be able to — the pipe or tubes are not climbable). All maintenance will be done on *terra firma*." When a wind turbine with a 10-foot diameter blade on an 80-foot tower needs to be inspected or serviced, you can lower the turbine and tower in less than half an hour, and have it back up and running in the same time — once you've completed the inspection or needed repairs. Taller towers and those supporting much larger turbines, a 24-foot diameter ARE 442, for example, may take a little longer to lower.

Although they're ideal for those who cringe at the idea of climbing a 80-to-120+-foot tower, tilt-up guyed towers do have a few drawbacks. For one, they have the largest footprint of all towers (Figure 6.8). This results from

IAN WOOFENDEN

several factors. First, tilt-up towers require four guy cables that anchor in the ground some distance from the tower. The larger footprint is also due to the need for room to lay the tower down. A considerable amount of room is also needed to operate a lift vehicle. As a general rule, the guy radius is about 35 to 80 percent of the tower height. The gin pole is typically 75 to 100 percent of the guy radius. The lay-down zone is equal to the length of the tower and the zone in which the lift vehicle operates is three times the height of the tower. For example, suppose you were installing a 120-foot guyed tilt-up tower. In this installation, the anchors

Fig. 6.15:
*Grip Hoist.
Ian's son, Zander,
demonstrates the
use of a grip hoist.
This device can be
used to raise and
lower tilt-up
towers.*

would be placed 42 to 72 feet from the base of the tower, depending on the manufacturer's recommendations. The lay-down zone would be 120 feet. The vehicle access zone would be about three to five times the height of the gin pole, or as much as 360 feet. This entire area would need to be free of trees and bushes so guy cables won't get caught in them as the tower is raised and lowered. The driving lane must also be free of obstructions.

Raising and lowering a tower isn't a job you should do alone. You'll need a couple of helpers to ensure that everything runs smoothly — for example, that the cables don't get tangled. And you'll need a truck, tractor, or some other lifting device.

Like all towers, things can go wrong with guyed tilt-up towers. They can get dropped while raising or lowering them and the wind turbine can be destroyed if it strikes the ground. Tow vehicles can slip. Accidents can also occur if the anchors are not correctly positioned or the guy cables get too tight while lowering or raising the tower.

Another disadvantage of tilt-up towers is that, except for the lattice towers sold for the Bergey Excel-S and Excel-R and ARE422 turbines, they are not climbable. This is a decided disadvantage if you are willing and able to climb and want to check out something simple on your turbine. Rather than climb the tower to repair it, you'll need to lower it and then raise it.

Even though Mick designed the first two tilt-up towers in the residential wind market many years ago, he prefers to climb towers than tilt them up and down. "Climbing puts you into a mental perspective where you are constantly on alert as you ascend or descend the tower, or work on the wind generator. You simply do not take risks at heights," notes Mick. Ground crews for a tilt-up tower operation, on the other hand, feel safe at ground level, and Mick has frequently seen their attention wander. This can be especially dangerous with inexperienced crews. "Just because you are on the ground doesn't mean you are safe," Mick warns. Should a distracted crew member lose control of the tower or should a problem occur with the tower, the consequences can be serious. The acceleration of a falling object — 32 ft/second^2 due to gravity — can result in a disaster *before* most people are aware that there is a problem. Anyone standing where a guy cable is headed on a tower that is no longer under the control of the winch or truck used for raising or lowering a tower ... well ... use your imagination. Think "egg slicer," and you'll understand the possibilities. After having installed numerous tilt-up towers and conducted multiple tilt-up tower installation workshops, Mick has come to the conclusion that the presumed safety advantage of a tilt-up tower is a myth. In fact, more people have had close calls with tilt-up towers than climbable towers. And one final point about tilt-up towers: although they offer key advantages, you'll never be able to enjoy the view from the top of your tower.

Guyless Tilt-Up Towers

In recent years, at least three US companies that we know of have introduced short, heavy-duty freestanding towers that can be lowered by hand, allowing inspection, maintenance and repair at ground level. These are described by the manufacturers as either hinged, tilt-up or tilt-down towers.

Independent Power Systems in Montana, for example, sells a 40-foot tower they describe as a guyless, tilt-down tower (GLT). It is designed for small wind turbines. The tower hinges 15 feet above the ground and can be raised and lowered by hand using an industrial hand winch. "An automatic disc brake in the winch prevents the tower from falling, and makes raising and lowering of the tower very safe," according to the manufacturer.

Unfortunately, the tallest guyless tilt-down towers we've found are 60 feet high. As noted in previous chapters, for best performance, most wind machines should be installed on 80- to 120+-foot towers, which raise machines into the strongest, smoothest winds. While a case could be made for installing short freestanding GLT towers in flat terrain (for example, grasslands) or along lake or ocean shorelines where obstructions to wind flow are minimal, even in perfectly flat terrain, companies such as Bergey Windpower, Abundant Renewable Energy, and Wind Turbine Industries recommend that towers be at least 80 feet high. Even on the treeless tundra, Ian would insist on a 60-foot minimum tower height.

Homebrew Towers

For those among us who like to do things themselves, the thought of making our own

tower might seem appealing. We've seen some ingenious designs and have great respect for independent-minded souls and applaud the creative efforts of do-it-yourselfers. When it comes to making your own tower, however, we urge extreme caution. "This is no place for lightweight construction or engineering guesswork," Ian notes in his article on towers in *Home Power* magazine. "If you are going to build your own tower, do careful research. Look at engineered towers and get a sense of the designs, as well as the size and quality of hardware used."

"When in doubt," he continues, "overbuild. Better yet, stick with engineered towers that are professionally designed for the job. To obtain permits, you may need an engineer's stamp on your plans anyway."

Tower Kits

If all this discussion of towers seems daunting, don't be dismayed. Several turbine manufacturers sell tower kits designed and engineered for their wind machines. Abundant Renewable Energy and Bergey, for instance, both provide guyed tilt-up tower kits for their turbines. Because steel pipe is heavy, it is expensive to ship long distances, so most kits include all of the materials you need *except* the pipe. You'll need to purchase the steel pipe locally.

You can also buy tower kits from tower manufacturers like Lake Michigan Wind & Sun, Mick's former company. Under the ownership of John Hippensteel, this company sells tower kits for turbines manufactured by a number of companies including ARE, Bergey, Proven and Southwest Windpower. The tower kits, designed by Mick when he owned the company, are far stouter than those of other companies and also more expensive.

When shopping, bear in mind that wind turbines often require adaptors, commonly referred to as stub towers, to fit onto commercially available towers. Stub towers consist of a short piece of pipe, with a flange that bolts onto the base of the turbine. Be sure to obtain an adaptor for your turbine. For best results, buy a hot-dipped galvanized adaptor. They last much longer than ungalvanized steel stub-towers.

When buying a tower, be sure you purchase one that's approved by the wind turbine manufacturer. Not doing so may negate the warranty! When shopping for a wind generator tower kit, also be aware that some kits may not include anchors, because the type of anchor needed for a tower varies from one location to the next, depending on the soil type, discussed shortly. In most cases, anchor foundations are constructed on site using concrete and rebar. You can purchase anchors from Southwest Windpower, Lake Michigan Wind & Sun, local wind energy suppliers/installers, or directly from turbine manufacturers.

While we are on the subject of tower kits, we want to warn readers against buying kits to mount towers and wind turbines on the roofs — or even against the walls — of buildings. Such installations are an unequivocally bad idea. Why?

First of all, roof-mounted turbines in most locations are too close to the ground, where wind speeds are slower. Lower wind speeds mean lower output. A 2007 study of roof-mounted wind turbines by Encraft, Ltd, a British consulting firm, showed that wind turbines mounted on homes and apartment buildings produced much less electricity than predicted by computer models. On residential structures, annual wind speeds were only about 5 to 7 miles per hour, well below the models' predicted wind speeds of 10 to 12 miles per hour. (The models obviously didn't account for ground drag and turbulence.)

Kilowatt-hour meters were also installed to monitor the output of the micro-turbines and small turbines in this study. The researchers compared energy generated by the urban turbines to the manufacturers' performance predictions. The researchers found that on average, wind turbines exported less than 0.5 kilowatt-hours a day. Some of the turbines generated less power than the inverters consumed and were therefore negative energy producers at their low wind speed sites. That is, these systems consumed more electricity than they generated.

The second reason rooftop installations are not a good idea is that buildings and associated ground clutter like trees create turbulence. Turbulence reduces the quality and quantity of wind. Lower wind speeds and turbulent winds reduce energy production. Therefore, roof-mounted wind turbines will produce significantly less electricity than turbines mounted on tall towers. More turbulence also means more wear and tear on the turbine, more costly replacements, and shorter turbine life.

Third, buildings are rarely designed and engineered to support the load and handle the vibrations produced by a wind turbine. These vibrations can cause structural damage to buildings. They may cause screws holding panels and inverters to come loose, causing damage. They may cause framing members to detach as well. Mick once saw a garage in which the rafter nails vibrated out by a roof-mounted wind turbine.

Fourth, noise conducted into the building can be disturbing if the building is occupied.

So, as nifty as this idea may seem, stay away from roof-top towers. Even in unoccupied buildings, the problems with roof mounts greatly outweigh the benefits of this approach.

Tower Anchors and Bases

When installing a tower, you'll need to install a strong base. If the tower is guyed, you'll also need to install anchors for the guy cables. All reputable manufacturers provide well-engineered plans for suitable foundations. Follow the plans very carefully. Don't cut corners to save money or time. Use common sense. Take a wind workshop or two *before* installing your wind machine or hire a professional to do the job.

Tower Base

Virtually all towers are mounted on concrete pads or piers reinforced with rebar. The depth required for a foundation, known as the critical depth, is the depth that prevents a foundation (and anchors) from being pried out of the ground by the force of the wind or pushed out by freeze-thaw cycles. The critical depth of a foundation depends on many factors, such as the type and the height of the tower, wind speed, depth of the frost line, and soil characteristics.

As noted earlier, freestanding towers require deeper and more robust foundations than guyed towers. The taller the tower, the stronger the foundation. The stronger the winds, the deeper and more robust the foundation. The deeper the frost line, the deeper the foundation. Foundations not installed below the frost line may be heaved out of the ground as the soil moisture freezes and expands. To determine the frost line in your area, check with your local building department or a knowledgeable builder in your area. "Then corroborate this with a local grave digger," Mick advises.

Soil characteristics also affect foundation design and construction. Some soils hold tower foundations in place better than others. Sandy soils, for instance, have less holding power than clay-rich and heavier soils, so tower foundations need to be much deeper than those in heavier clay-rich or rocky soils. For advice on critical tower depth, contact the turbine and tower manufacturers.

As noted earlier, freestanding towers require the most concrete and steel. In a lattice tower, each leg of the tower is attached to a steel anchor that's embedded in a steel-reinforced concrete pier (Figure 6.16). The base of the pier is often wider than the column above it, creating a footing that distributes the weight of the tower. Alternatively, piers may be attached to an underlying concrete foundation that joins all three piers together (Figure 6.5). The pad

Fig. 6.16: *Most wind turbine towers rest on a solid concrete base. This photo shows the concrete pier for each leg of a freestanding lattice tower.*

acts like a footing and provides even greater resistance to the wind.

Free-standing monopole towers require a single deep, robust steel-reinforced pad located well below the frost line. Foundations for guyed and tilt-up towers are far simpler than those for monopoles. They require much less concrete and rebar for reasons explained earlier.

The most common concrete foundation for these towers consists of a single square or round concrete pad. The thickness of the pad varies from one site to the next, depending on climate (depth of the frost line), soil strength, wind speed and tower height. Thin pads are used in warm climates. Deep, thick pads or other types of foundation are used in colder climates.

Engineered plans are typically provided in installation manuals that come with turbines. Even so, when installing a wind generator tower, it's a good idea to consult local building codes, soil engineers or local excavators to develop a tower foundation that's affordable yet sufficiently strong.

Concrete pads for tilt-up and guyed towers require hardware to attach the tower. To secure a guyed pole tower, for instance, installers embed anchor studs, 26-inch-long bolts threaded on both ends, in the pad. They are bolted into a one-half-inch steel anchor plate, shown in Figure 6.17, which is embedded in the concrete foundation, and attached to a flange plate on the base of the tower.

Other foundations use longer J-bolt anchors embedded in the concrete pad (Figure 6.18). Be sure to follow the manufacturer's recommendation.

For maximum strength, many professionals recommend 28 days for concrete to cure prior to installation of the tower. Whatever you do, do not place a tower on a concrete pad before this time! Inadequately cured concrete weakens the concrete. So don't rush it! Your tower and wind machine rely on a sturdy, well-cured

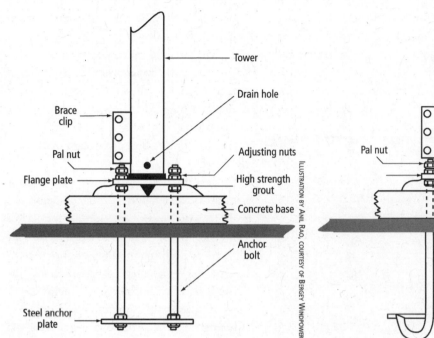

Tower

Drain hole

Brace clip

Pal nut

Adjusting nuts

Flange plate

High strength grout

Concrete base

Anchor bolt

Steel anchor plate

ILLUSTRATION BY ANIL RAO, COURTESY OF BERGEY WINDPOWER

Fig. 6.17: *This tower is attached by anchor bolts. The anchor bolts connect the flange plate to a steel plate embedded in the concrete foundation.*

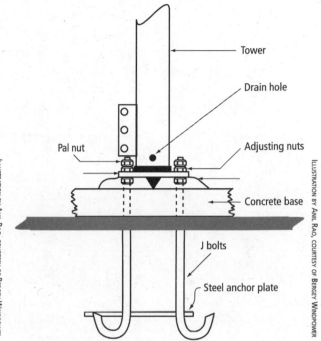

Tower

Drain hole

Pal nut

Adjusting nuts

Concrete base

J bolts

Steel anchor plate

ILLUSTRATION BY ANIL RAO, COURTESY OF BERGEY WINDPOWER

Fig. 6.18: *In this tower, J bolts anchored in the concrete pad are attached to a flange plate at the base of the tower, thus securing the tower to the deep steel-reinforced concrete anchor.*

foundation. Remember, too, that cold weather slows the curing process. More time may be required in such conditions.

Anchors

In addition to a concrete base, fixed guyed towers and guyed tilt-up towers also require anchors to attach the guy cables. Anchor options for wind generator towers are many and varied. For most wind turbines, concrete anchors are the best choice. A concrete anchor is created by digging a deep hole, installing rebar, and then pouring concrete into the hole. A screw-in auger, anchor, or angle steel is then embedded into the hole and angled towards the tower as per manufacturer instructions. Guy cables are attached to these structures (Figure 6.19). Once the concrete has cured, the hole is filled in. Be sure to pour anchors below the frost line and follow the advice of the tower manufacturer or supplier.

Augers and duckbills are additional options, but are typically only suitable for very small wind turbines, the micro-turbines installed on

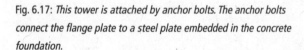

short towers. (As we've stated many times, we frown on the use of short towers.)

Augers are manually screwed into the ground using a piece of pipe or a steel rod, known as a cheater bar, which makes it easier to turn the auger. (They provide leverage that reduces the force you must apply.) Augers should be screwed in at an angle tilted toward the tower as specified by the manufacturer (Figure 6.20).

Longer heavy-duty augers, measuring four feet or more are appropriate for short wind generator towers (under 60 feet). They are available through turbine and tower manufacturers such as Southwest Wind Power or Lake Michigan Wind & Sun or through utility supply houses.

Augers are easy to install, fairly inexpensive, and can also be removed if you make a mistake in placement. Augers are suitable for clay or loamy soil, but can be difficult to install in rocky soil, and do not hold in sandy or gravely soil.

Another option for short towers is the duckbill. A duckbill is a small metal device that's attached to the guy cable and driven at an angle into the ground using a steel rod (Figure 6-21). Once it's been driven to the desired depth, the installer tugs on the duckbill. This turns it slightly perpendicular to the surface. Like a deadman, it holds the cable securely in the ground.

Fig. 6.19: *Guy cables attach to the concrete anchor via a metal plate embedded in the anchor. Turnbuckles attach to the plate and the guy cables allow them to be properly tensioned.*

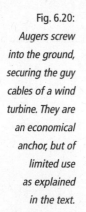

Fig. 6.20: *Augers screw into the ground, securing the guy cables of a wind turbine. They are an economical anchor, but of limited use as explained in the text.*

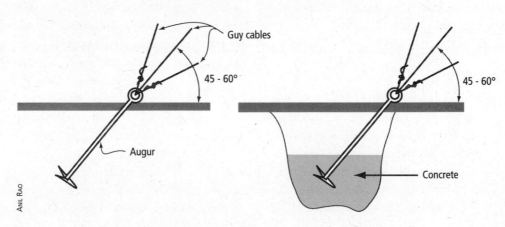

Duckbills are very inexpensive and easy to install. However, they are only suitable for small turbines on short towers in rocky soils. They're a nightmare to remove and the guy cables run underground where they can rust.

For sites with lots of large rock or on solid bedrock, expansion bolts or large eyebolts can be cemented with epoxy into the rock. Expansion bolts can be used in harder rock such as granite and for softer rocks, such as the harder grades of sandstone, although longer bolts are necessary.

To insert an eyebolt, a hole must first be drilled into the rock. The bolt is then placed into the hole with an attachment hanger to connect to the guy cable. Epoxy is injected around the bolt. We recommend that you consult the tower manufacturer or a company familiar with rock anchors and rock structure before you install a wind generator tower using anchor bolts.

As you can see, there are a variety of anchor options available for wind generator towers, however, most of them are unsuitable for taller towers — 80 to 120+ feet. For tall towers, the kind you should be using to install your wind machine, concrete anchors are your best bet. They're stronger, long-lasting, fairly inexpensive, and easy to construct.

Safety First

Wind energy is a great source of electricity. Unlike PV modules in a solar electric system that provide years of maintenance-free service, a wind turbine requires routine inspection and maintenance and occasional repair. This, in

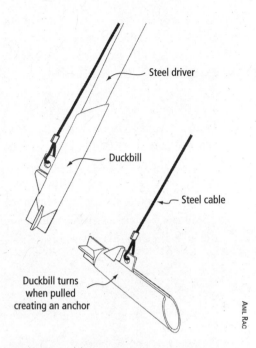

Steel driver

Duckbill

Steel cable

Duckbill turns when pulled creating an anchor

Anil Rao

Fig. 6.21: *A duckbill is driven into the ground using a steel rod and hammer (top figure). When it's deep enough, the installer tugs on the cable, which causes the duckbill to turn, securely locking the cable in the ground (right figure).*

turn, requires periodic ascent of the tower or lowering and raising of the tower. Both activities pose some risk.

Installing a tower is also a risky business — even more so if it is your first experience. Safety of workers and onlookers should be an installer's primary concern during an installation. Moreover, nearby equipment and structures could be damaged if a tower collapses or topples during installation. And, lest we forget, wind turbines can also be damaged if a tower collapses or falls to the ground during installation. At the very least, the blades are very likely going to break in a fall. The generator could also be damaged. If all parties remain aware of the dangers and observe all of the safety rules, risk can be minimized.

With this in mind, we offer some safety tips for tower installation, reprinted below with permission and some modification from Bergey Windpower's XL.1 installation manual.

1. Individuals not directly involved in raising the tower should stay clear of the work area.
2. All individuals on or in the vicinity of the tower should wear OSHA-approved hard-hats. (Hard hats are meant to protect your head from bumping against something, not to protect your head from something dropped from a tower.)

Fig. 6.22: *Lightweight and easy to carry, this canvas bucket is used to transport tools up a tower and hold tools while service personnel are working on a wind turbine. Never carry a plastic bucket, as the handle could come off.*

DAN CHIRAS

3. Tower work should be performed by — or should be under the supervision of — a trained installer.
4. Towers should never be constructed near utility lines. If any portion of the tower or equipment comes into contact with them, serious injury or death may result.
5. People working on a raised tower should wear a safety harness, and a tool belt, and use an anti-fall climbing cable and safety device (discussed earlier).
6. Tools should not be carried in one's hands when climbing a tower. Use a hoistable tool bucket with a closure (Figure 6.22).
7. Never stand or work directly below someone who is working on the tower.
8. Never work on the tower alone.
9. Never climb the tower unless the wind turbine is shut down and the alternator is secured with a brake or by a short circuit. The best location for the short circuit switch or jumpers is in the tower-base disconnect switch box.
10. Never drop or throw anything from the tower. If you drop lines when working on a tower, warn the ground crew and be sure to receive visual or auditory feedback that they've heard your warning. Be sure they have cleared the area.
11. Stay off the tower during bad weather or impending bad weather, including thunderstorms and high winds. Stay off the tower after ice storms.
12. If you see bad weather approaching, climb down immediately and suspend activities until the storm has passed.

As a final note on safety, beware of funky tower configurations — inexpensive, light-duty, highly questionable towers that could be characterized as "back-of-the-envelope" designs. You may, for example, hear of people installing a wind machine atop a tree to save on tower costs. Don't do it. It's more trouble than it is worth, difficult to pull off, and fraught with problems.

Siting a Tower

As you know by now, to generate electricity economically and in sufficient quantity to meet your needs, you need to place a wind turbine in a good wind site. For optimum electrical output, a wind machine must be mounted well above the ground, out of the way of ground clutter, in the strongest, smoothest winds. Careful siting and proper tower height combat the two biggest enemies of wind systems, ground drag and turbulence, discussed at length in Chapter 2. Let's begin with proper siting.

Proper Siting of a Wind Machine

When siting a wind generator — that is, determining a location that will allow the wind turbine to perform optimally — remember at all times that your main goal is to situate the tower out of the turbulence bubble created by ground clutter.

Wind site assessors begin this process by determining the prevailing wind direction at a site. Although winds blow in different directions at different times of the year, even within the same day, they'll typically blow from one or two directions predominantly over the course

of the year. In many places in North America, for example, the winds come predominantly from the southwest — thanks to the Coriolis effect described in Chapter 2. In some locations, they blow from the northwest in the winter and the southwest in the summer. How do you determine the predominant wind direction?

One way is to ask long-time residents for advice. Especially helpful are farmers, who work outdoors and hence are familiar with wind patterns. Another, potentially more reliable way

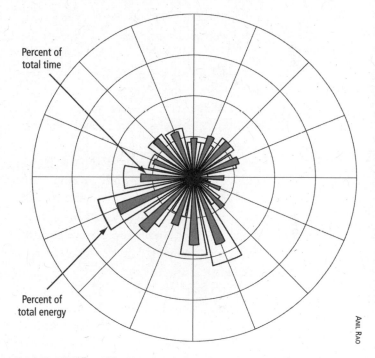

Percent of total time

Percent of total energy

Anil Rao

Fig. 6.23: *Wind Rose. This unique graph shows how often winds blow from various directions and the percent of total energy in the wind from various directions. The wider white bars represent the percent of total energy from different directions and the narrower, shaded bars illustrate the percent of total time from each of the sixteen different direction sectors.*

to determine predominant wind direction is to contact a nearby airport. They may be able to provide you with a wind rose, a graphical representation of wind direction (Figure 6.23). In a wind rose, shaded bars indicate how frequently the wind blows from a particular direction. The longer the line, the greater the total time from a particular direction. In the example shown in Figure 6.23, the winds blow predominantly from the southwest. A wind rose also indicates the percentage of total wind energy from each direction, which is very helpful. In this example, most of the energy comes from west and southwesterly and south and slightly southeasterly winds. You can also find data on wind direction at the NASA Surface Meteorology site (Chapter 4).

In an open site, with little ground clutter, a wind turbine can be located almost anywhere — so long as the entire rotor is mounted 30 feet above the tallest obstacle within a 500-foot radius and you've taken into account future tree growth, if trees are the tallest objects. Unfortunately, very few of us live on ideal sites. There's almost always some major obstacles that make siting difficult. What do you do in such instances?

Once you have determined the prevailing wind direction, look for a location for the tower upwind of major obstacles. Although winds will shift so that upwind becomes downwind, your wind turbine and tower will be situated so that it takes advantage of the strongest prevailing winds. You still need to obey the 30/500-foot rule. This is frequently most easily accomplished by installing your tower on one of the highest spots on your property.

When siting a wind turbine, it is also a good idea to minimize wire runs from the turbine to the controller and inverter to reduce line loss. As a rule, the higher the wind turbine's voltage, the farther it can be sited from the point of use. When installing a turbine, contact the manufacturer or an experienced installer to discuss the length of wire runs. All manufacturers should supply a chart showing recommended wire size based on turbine type and length of wire run.

Tower Height Considerations

Once you've identified the best site, it's time to determine optimum tower height. As noted in previous chapters, for a wind system to perform optimally — to produce as much electrical energy at the lowest cost possible at a site — it needs to be on a tall tower. Once again, the rule of thumb is that the *entire* rotor should be *at least* 30 feet above the tallest obstacle within a radius of 500 feet.

Consider an example. Suppose you were going to mount a wind machine on a farm with a 45-foot silo 250 feet upwind of the proposed tower site. Suppose the blades of the wind turbine are ten feet long. To determine the minimum acceptable tower height, add the silo height (45 feet) to the blade length (10 feet) to the 30-foot height allowance. The minimum tower height is 85 feet. Since towers come in 10-foot sections, round up to at least 90 feet. Remember, however, that 85 feet is the minimum acceptable tower height. Because output increases on taller towers, we strongly recommend that you also analyze the cost and output of the same turbine on a 100- and 120-foot towers.

When calculating minimum tower height, don't forget to take tree growth into account. If the trees on your property will grow 20 feet in the next 20 to 30 years, the life expectancy of a wind system, add that to the tower height for the best long-term performance. Anything less will reduce the performance of the wind generator and reduce its life. For example, suppose you were going to install this wind turbine within 500 feet of a 50-foot tree whose mature height is 75 feet. In this case, the tower should be at least 115 feet. That's 75 feet for the tree height added to the blade length (10 feet) plus the 30-foot rule. Since the tower you want in this example comes in 20-foot sections, round up to at least 120 feet. A turbine on a 140-foot tower would perform even better! (To learn two ways to estimate the height of trees and buildings, see the accompanying box.)

If your site is within a quarter of a mile from a forest or good-sized wood lot, the top of the nearby tree line is the height you want to exceed. Mount the wind turbine as above, using the tree line as the height you must exceed. And remember that the trees will very likely grow.

To determine tree growth, you can check out the mature height of the trees in the area. If the trees on your property are 50 feet tall and top off at 65 feet, use this number. Anything less will reduce the performance and life expectancy of your system. You can find information on mature tree height in field guides on trees and on the Internet.

While we're on the subject of tree growth, remember that tree growth varies in response to water supplies. In wetter climates, trees grow more rapidly than in drier areas. Take this into consideration.

If you are installing a wind turbine in an area with more than 50 percent tree cover, the effective ground level is two thirds of the tree height. If trees are 60 feet high, for instance, the effective ground level, known as the displacement height, is 40 feet. A 100-foot equivalent tower would, therefore, need to be 140 feet high to take into account the trees.

In the literature on wind energy systems and in product literature, you will see statements contradicting our advice on turbine height. It is not unusual, for instance, to read that a wind machine should be placed at least 20 feet above obstructions within 250 feet.

As far as we're concerned, the acceptable numbers are 30 feet and 500 feet. Bear in mind, too, that this is the *minimum* acceptable tower height. Taller towers are generally better. Savvy wind energy installers generously exceed the basic rule of thumb we've presented, and see increased performance because of it. It usually costs very little to increase tower height by another 20 to 40 feet and the return on this small investment is usually quite impressive. We don't know anyone who has installed a wind turbine who says, "I wish I'd bought a shorter tower." But we know lots of people who wish they had purchased a taller tower.

Figure 6.25 shows a wind turbine that was installed in the early 1980s two miles from Mick's home. At the time of the installation,

How High?

Determining the height of a tower seems pretty straightforward until you have to do it. The first challenge you'll face is determining the actual height of nearby objects, such as trees or barns. How do wind site assessors determine the height of ground clutter?

One way, shown in Figure 6.24a, is to place a stake (a metal fence post, for instance) next to the object you want to measure. On a sunny day, measure the height of the stake and then measure its shadow. Then measure the shadow of the object under question.

The height of the object can be determined by ratios using the equation: $\frac{H_1}{H_2} = \frac{SL_1}{SL_2}$. H1 is the unknown height and H2 is the height of the fence post. SL1 is the length of the shadow of the object you are trying to measure. SL2 is the length of the shadow of the fence post. To solve for H1, you'll need to rearrange the equation: $H_1 = \frac{(SL_1)(H_2)}{SL_2}$. Note that the ground around both the tree and the fence post must be level for this method to be accurate.

Consider an example. Let's assume that the fence post is four feet high and the shadow it casts is two feet long. The shadow cast by the tree or building is 14 feet. How high is the tree? As illustrated, in Figure 6.24a, you begin by setting up ratios: $\frac{x}{4} = \frac{14}{2}$, then solve for x: $x = (4 \times \frac{14}{2}) = 28$ feet. Another simple method is explained in Figure 6.24b.

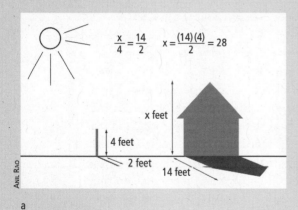

$$\frac{x}{4} = \frac{14}{2} \qquad x = \frac{(14)(4)}{2} = 28$$

x feet

4 feet

2 feet

14 feet

ANIL RAO

a

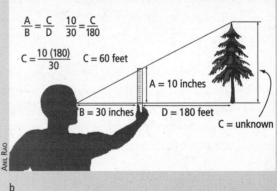

$$\frac{A}{B} = \frac{C}{D} \qquad \frac{10}{30} = \frac{C}{180}$$

$$C = \frac{10\,(180)}{30} \qquad C = 60 \text{ feet}$$

A = 10 inches

B = 30 inches   D = 180 feet

C = unknown

ANIL RAO

b

Fig. 6.24: *Measuring Height. (a) Driving a fence post or some other object of known length into the ground next to an object of unknown height and comparing the length of the shadows, allows one to calculate the height of an object. (b) Another method for determining height is shown here. In this method, you'll be solving for C, the height of the tree. Measure the distance from the tree (D). Measure the distance from your eye to the ruler in your hand. This is B. Measure the height of the object in inches on the ruler. This is A. Then set up a ratio equivalence as follows: $\frac{A}{B} = \frac{C}{D}$. The rest of the math is shown in the figure.*

the trees were well below the height of the house. The installers, Mick and the owner were greenhorns and didn't take tree growth into account. The result?

Today, the wind machine is hidden from the wind. The blades of the wind machine are so sheltered that they only occasionally spin, and the machine produces no electricity. It's become an expensive lawn ornament. So why doesn't the owner just raise the tower?

Unfortunately, it is hard to "grow" towers, unless you've planned and engineered for it in advance. Tower anchors and guy radii are specified for a specific height. If you decide to go higher later, you usually will need to start over from scratch — which can be expensive.

Tall Tower Economics: Overcoming the Small-Turbines-on-Short-Tower Myth

When you talk to professional wind system installers, you may hear statements to the effect that it doesn't make sense to mount a smaller turbine, for example, one with a seven- or eight-foot diameter rotor, on a tall tower.

We strenuously object to such statements. Tower height should be determined by the height of obstructions in the area, not the size of the wind turbine or the towers a manufacturer or dealer sells. A 50-foot tower slightly downwind from 65-foot-high treeline isn't going to produce much electricity. Moreover, the turbine will produce even less electricity as the trees grow over the 20- to 30-year life of the wind system. Remember: energy output and the economics of the wind system are both proportional to V_3 (the cube of the wind speed).

Fig. 6.25: *This wind machine no longer produces any energy; the trees around it have grown taller than the turbine and successfully block virtually all wind.*

Although it is sometimes hard to justify a tall tower for a small turbine, that doesn't mean that the right decision is a short tower. The right decision is to invest enough in your tower to make the most of your turbine's potential — or choose another renewable energy system.

If you are thinking about installing a smaller wind generator, but are nervous about the cost of a taller tower, we recommend that you calculate how much more the tower will cost and how much more electricity the turbine will produce on a taller tower. In our experience, installing a taller tower always results in the production of more electricity. Even though it will always cost more money, the important question to ask is whether the increased tower height is justified economically by the increase in electrical production. In most cases, it is. Consider an example.

Let's assume that we are interested in installing a wind generator with a 23-foot diameter rotor, and that the bare minimum tower that will work at the site is 60 feet. We'll use the average wind speeds at the Midwest Renewable Energy Association site in Custer, Wisconsin, a fairly average site in North America (Table 6.3). Using a wind turbine output calculator, we can determine the annual kilowatt-hour (kWh) output of this wind turbine at the three tower heights.

After estimating electrical output, the next thing we need to do is to determine the costs of the material and labor required for each tower height. Table 6.4 shows the costs.

As you can see, it costs very little additional money to install a slightly taller tower. Even so, we are still left with the question: is it worth it?

As Table 6.5 shows, the cost of increasing the tower from 60 feet to 80 feet only adds 2.4 percent to the overall system cost, but increases energy output by 226 percent (see columns 4 and 5). The cost of going from 60 to 100 feet increases the system cost by 5.5 percent but increases the energy output by 344 percent. Both represent incredible returns on investment.

Mick's advice to individuals interested in wind energy is to determine the minimum tower height required for the site first. Then calculate the costs and benefits of a taller tower.

If your calculations show that a taller tower is worth it, and you can afford one, go for it. However, cutting costs by cutting tower height is foolish. It is akin to installing a PV system on your home, but putting it in a shady location because it's cheaper or more convenient to install it there.

Some might ask, would it make sense to install a larger turbine on a small tower? Once again, an economic analysis shows that this is a foolish choice. To see why, let's compare a 2.5 kilowatt-hour turbine at 42 feet to a 1 kilowatt-hour turbine at 84 feet. Both options produce about the same amount of energy. The larger turbine at 42 feet, however, cost $19,500 to install in large part because of the higher cost of the larger turbine. The smaller turbine at 84 feet, even with a taller tower, only cost $14,700. (For details, see Mick's article, "Tall Tower Economics" in *Windletter* (No 25, Issue 2).

If that doesn't work, how about installing several small turbines on short towers?

Once again, the economics bode poorly for this strategy. In a home that requires 9,000 kilowatt-hours per year, a single turbine on a 100-foot tower that would meet this demand

Mick Sagrillo, "Incremental Tower Costs vs. Incremental Energy" published in *Windletter*

Table 6.3		
Wind Turbine Output at Different Heights		
Tower Height (feet)	**Average Wind Speed (mph)**	**Electrical Production (kWh/year)**
60	7.3	2,709
80	9.3	6,136
100	10.7	9,338

Table 6.4
Estimated Cost of Installing Towers

Expense	60 foot tower	80 foot tower	100 foot tower
Excavation for foundation and backfilling	$800	$800	$800
Firm up foundation, pour concrete	$960	$960	$960
Concrete and rebar	$750	$750	$750
Wind turbine and inverter	$24,750	$24,750	$24,750
Shipping of wind turbine and inverter	$800	$800	$800
Tower	$7,400	$8,100	$9,200
Shipping of tower	$900	$950	$1,000
Tower wiring	$800	$860	$930
Wiring and electrical supplies	$1,800	$1,800	$1,800
Electrical labor	$1,200	$1,200	$1,200
Crane	$750	$750	$750
Installation labor for tower, turbine, commissioning, mileage and expenses	$5,000	$5,300	$5,500
Sales Tax (6%)	$2,755	$2,821	$2,906
Total	**$48,665**	**$49,841**	**$51,346**

Table 6.5
Incremental Cost vs. Incremental Energy Output*

Tower Height (Feet)	Wind Speed (mph)	kWh/year	System Cost	Incremental Cost from 60 feet	Incremental energy output from 60 feet
60	7.3	2,709	$48,665	--	--
80	9.3	6,136	$49,841	$1,176 or 2.4%	226%
100	10.7	9,338	$51,346	$2,681 or 5.5%	344%

*This analysis is specific to the particular site used in this example and the particular installation, and cannot be assumed to be valid for other situations or installations. However, the technique for doing the analysis for the cost effectiveness of incremental tower heights is the same for any equipment or site.

would cost approximately $51,350. A smaller turbine at 80 feet and another at 60 feet that together would produce 9,000 kilowatt-hours per year, cost nearly twice as much, about $98,500. Three small turbines at 60 feet that together produce 9,000 kilowatt-hours per year would cost nearly three times as much, or nearly $146,000.

Any way you look at it, a single tall tower is the best investment. If for some reason you don't want to invest in a tall tower or don't want a tall tower on your site, we recommend that you consider installing a PV system instead, provided your solar resource is adequate. We can't emphasize enough that tower height determines how much "fuel" you collect from the wind, and subsequently how much electricity you will generate. The tower height is dictated by the site, *not* what the manufacturer or dealer has to offer.

Aircraft Safety and the FAA

Another factor to consider when installing a wind turbine is its impact on aviation and the possible need to file for a permit from the Federal Aviation Administration (FAA).

Fortunately, the FAA requires a permit application under only two conditions. The first is if the height of the turbine and tower is over 200 feet, which is extremely rare for a small wind turbine. The 200-foot requirement on a wind turbine and tower refers to the total height of the turbine and tower with a blade in an upright position. To remain below the 200-foot limit, then, a wind turbine with a 30-foot rotor diameter (15-foot blades) would have

to be mounted on a 185-foot tower (185-foot tower plus 15-foot blade = 200 feet). Most home-sized wind towers do not exceed the 120-foot mark.

The tallest towers offered by manufacturers and dealers as "off the shelf" items are in the 120- to 140-foot range, well below the 200-foot mark that could trigger a review by the FAA. In instances where tower heights exceed the 200-foot mark, the FAA may require top-of-tower lighting to warn pilots.

The second situation that would require a FAA review and permit is based on the proximity of a wind generator tower to an airport — specifically a "public use" or military airport. Permit applications are required if you are installing a turbine within 3.79 miles of an active public use or military airport if the runway is longer than 3,200 feet. If the runway is less than 3,200 feet long, you'll need to file an application if a turbine is being installed within 1.9 miles of the runway. In either case, the FAA will determine the height of the tower you can install. They may also require top-of-tower lighting. Bear in mind that this requirement does not pertain to private landing strips with no public access, airfields not shown on FAA maps, or landing strips that are not in active use.

An FAA application takes at least several months to process, so plan ahead. If you are in doubt about your distance from an airport, or the status of the airport, it is always best to submit an application. If you're hiring a professional installer, he or she can advise you on this matter. The application form is #7460

(Notice of Proposed Construction or Alteration). Instructions are available at the FAA website forms.faa.gov/forms/faa7460-1.pdf. Addresses of the regional FAA offices where applications should be mailed are listed on the instructions.

If your project is outside the FAA guidelines, do not submit an application. In most cases, Mick notes, "the FAA does not want to be bothered with the permitting of a homeowner's wind system."

Protecting Against Lightning

When installing a wind turbine, it is important — indeed essential — to include lightning protection, specifically ground rods and lightning arrestors.

As air masses move over the surface of the Earth, a static charge builds up between the Earth and the atmosphere, essentially creating a giant capacitor storing this static charge. At some point the static electricity has to equalize. It does so between 2,000 and 3,000 times a minute somewhere on the planet. This is the phenomenon called lightning.

Although lightning is not necessarily attracted to tall metal objects such as a wind generator tower, bleeding off the static charge by installing ground rods will reduce the likelihood of a strike. Ground rods are eight-foot long copper-coated rods. They are driven into the ground at the base of the wind turbine tower.

For a three-legged freestanding tower, you'll need three ground rods, one for each leg. For a guyed tower, you'll need four ground rods, one for the tower and one for each of the three guy

---

FAA Applications and Zoning Hearings

The need for an FAA application sometimes arises during zoning hearings for small-scale wind systems in an attempt to delay the installation of a turbine and tower. However, this is a matter between the applicant and the Federal Aviation Administration. Zoning officials should not use the FAA application, if any, to delay the installation of a wind turbine and tower, or as an added hurdle to dissuade homeowners from installing wind turbines.

---

anchors. For a guyed tilt-up tower, you'll need five ground rods; one for the tower and one for each of the four guy cables.

Grounding a structure's static charge does minimize lightning strikes, but it cannot guarantee that lightning will not strike a tower. Backup is needed in the form of lightning arrestors and surge arrestors.

Lightning arrestors are known in the trade as silicone oxide varistors or SOVs. An SOV is a spark gap arrestor. Inside the SOV is a device that looks like a spark plug. It is embedded in white sand. One end of the "spark plug" is attached to the wind turbine's wiring; the other end is connected to the ground rod at the base of the tower. A spark jumps the gap when a certain voltage is reached. In theory, then, if lightning strikes a tower and travels down the wiring towards the battery bank or inverter in your house, the SOV shunts that lightning strike to ground. Mick stresses "in theory" because, like ground rods, there's no guarantee that a lightning arrestor will work every time.

Sometimes lightning doesn't follow human rules and goes where it wants to.

SOVs should be installed on the electrical wire running down the tower as well as the utility wiring for grid-tied systems to protect against lightning strikes in utility lines. In reality, lightning strikes are orders of magnitude more frequent on the utility grid than on properly grounded towers.

While protecting against direct lightning strikes is important, nearby strikes pose the gravest danger to a wind system. When lightning strikes the ground near a home, for instance, it creates an electrical current in the atmosphere and/or the ground. This current creates a bull's eye voltage wave pattern that resembles ripples in a quiet pond after a pebble is dropped into it. If one of these waves crosses a conductor like a wind generator tower or buried wires, an electrical current will be created (induced) in that conductor. This current can fry sensitive electronic components of a wind system.

To protect against this problem, installers install metal oxide varistors, or MOVs, on the tower wiring — that is, the electric wires that run from the turbine to the base of the tower. (MOVs are used in computer surge arrestors to offer the very same protection to the sensitive electronics in our homes from surges that occur on utility or telephone lines.)

Will these measures guarantee that you will never be struck by lightning?

Maybe.

Lightning, which travels miles through the atmosphere, which is a poor conductor, will pretty much do as it pleases. Ground rods, SOVs and MOVs may help but they aren't fail-safe. By installing these protective devices to a system, you have installed all of the precautionary devices offered by our best understanding of the science of lightning. You have done all that a prudent homeowner can do. If your system sustains damage, you should receive compensation from the insurance company.

CHAPTER 7

UNDERSTANDING BATTERIES

Batteries are the unsung heroes of off-grid wind energy systems. Even though wind generators work hard and endure extreme weather, the batteries in off-grid systems operate day in and day out 365 days a year for years — up to 10 years, give or take a little. These brutes of the renewable energy field are not so brutish as you might think. They require considerable pampering. You can't simply plop them in a battery room, wire them into your system, and then go about your business expecting them to perform at 100 percent of their capacity for the lifetime of your system. You'll need to watch over them carefully and tend to their needs. If you don't, they'll die young, many years before their time.

Because batteries are so important to off-grid renewable energy systems and require so much attention, it is essential that readers thinking about installing an off-grid system or a grid-connected system with battery backup understand batteries, and understand them extremely well. Doing so will pay huge dividends.

This chapter will help readers develop a solid understanding of batteries. It will help them select the proper batteries for a renewable energy system and will provide guidance on installation and maintenance to optimize their performance.

What Types of Batteries Work Best?

Batteries are a mystery to many people. How they work, why they fail, and what makes one type different from another are topics that boggle the mind. Fortunately, the batteries used in most off-grid renewable energy systems are pretty much the same: deep-cycle flooded lead-acid batteries. These batteries can be charged and discharged (cycled) hundreds of times before they wear out.

Lead-acid batteries used in most renewable energy systems contain three 2-volt cells — that is, three distinct 2-volt compartments. Inside, the individual cells are connected (wired in series) so that they produce 6-volt electricity. Inside each cell in a flooded lead-acid battery is a series of thick parallel lead plates, as shown in Figure 7.1. Each cell is filled with sulfuric acid (hence the term "flooded"). A partition wall separates each cell, so that fluid cannot flow from one cell to the next. The cells are encased within a heavy-duty plastic case.

As illustrated in Figure 7.1, two types of plates are found inside batteries: positive and negative. The positive plates connect to a positive metal post or terminal. The negative

plates connect internally to a negative post. The posts allow electricity to flow into and out of batteries.

The positive plates of lead-acid batteries are made from lead dioxide (PbO_2). The negative plates are made from pure lead. Sulfuric acid fills the spaces between the plates and is referred to as the electrolyte.

How Lead-Acid Batteries Work

Although battery chemistry can be a bit daunting to those who have never studied the subject or had it explained well, it's important that all users understand the reactions.

To begin, like all other types of batteries, lead-acid batteries convert electrical energy into chemical energy when they are being charged. When discharging, that is, giving off electricity, chemical energy is converted back into electricity. To understand how, let's start with the chemical reactions that take place during battery discharge.

As shown in Figure 7.2, when electricity is drawn from a lead-acid battery, sulfuric acid reacts chemically with the lead of the negative plates (anode). This chemical reaction, shown in the top reaction, yields electrons, tiny negatively charged particles. They flow out of the battery at the negative terminal. During this reaction, lead on the surface of the negative plates is converted to lead sulfate, creating tiny lead sulfate crystals. When a battery is discharging, sulfuric acid also reacts with the lead dioxide of the positive plates, resulting in the formation of tiny lead sulfate crystals on them as well. This is shown in the same equation in

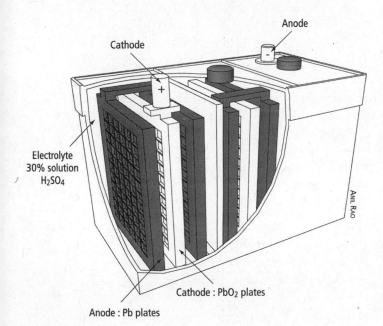

Anode

Cathode

Electrolyte
30% solution
H_2SO_4

Anil Rao

Cathode : PbO_2 plates

Anode : Pb plates

$$Pb(s) + HSO_4^-(aq) \longrightarrow PbSO_4(s) + H^+(aq) + 2e^- \qquad \text{anode}$$

$$PbO_2(s) + HSO_4^-(aq) + 3H^+(aq) + 2e^- \longrightarrow PbSO_4(s) + 2H_2O(1) \qquad \text{cathode}$$

Fig. 7.2: Chemical Reactions in a Lead-Acid Battery during discharge.

Figure 7.2. Discharging a battery not only creates lead sulfate crystals on the positive and negative plates, it depletes the amount of sulfuric acid in the battery. When a battery is recharged, the chemical reactions run in reverse. During recharging, then, lead sulfate crystals on the surface of the positive and negative plates are broken down and sulfuric acid is regenerated. The positive and negative plates are restored. Electrons are stored in the chemicals in the system.

Although the chemistry of lead-acid batteries is a bit complicated, it is important to remember that this system — and other battery systems as well — works because electrons can be stored in the chemicals within the battery when a battery is charged. The stored electrons can be drawn out by reversing the chemical reactions. Through this reversible chemical reaction, the battery is acting as a "charge pump," moving electrical charges through a circuit on demand.

Will Any Lead-Acid Battery Work?

Lead-acid batteries come in many varieties, each one designed for a specific application. Car batteries, for example, are designed and manufactured for use in cars, light trucks and vans. Marine batteries are designed for boats; golf cart batteries are designed for use in golf

A Word of Warning

Sulfuric acid is a *very* strong acid. In fact, it is one of the strongest acids known to science. In flooded lead-acid batteries, sulfuric acid is diluted to 30 percent. Although diluted, it is still a chemical to treat with great respect — it can burn your skin and eyes and eat through clothing like a hungry moth.

Fig. 7.3: Deep-Cycle Lead-Acid Batteries. These batteries contain thick lead plates and are used in many battery-based renewable energy systems. The thick plates permit deep cycling so long as the batteries are recharged soon after each deep discharge.

SURRETTE BATTERY COMPANY LTD.

carts. Forklift batteries are used to power electric fork lifts.

For off-grid systems, your best bet is a deep-cycle flooded lead-acid battery like those made by Trojan, Rolls, Deka and others, (Figure 7.3). Forklift batteries also work well, as do golf cart batteries. Both, however, have limitations we'll discuss shortly. Whatever you do, stay away from car batteries and marine deep-cycle batteries.

Why Won't Car Batteries Work in a Renewable Energy System?

Unlike deep-cycle batteries, automobile batteries contain very thin lead plates. They provide a very large surface area, in turn, that permits a high-amp discharge (200 or more amps) needed to run the starter motor used to start a car or truck.

Although car batteries produce a high-amp surge, this burst of electricity only needs to last a few seconds. Consequently, automobile batteries rarely discharge more than a few percent of their capacity — the total stored electricity — at any one time. Shallow discharges cause very little damage to the lead plates. Once the engine starts, the alternator quickly replenishes the energy drawn from the battery and the battery is then sidelined. That's because, once the engine starts, the alternator produces electricity needed to power the car's many electrical loads.

Although car-starting batteries perform admirably in our beloved automobiles, they fail miserably in renewable energy systems because they can't withstand the repeated deep discharge that is required of them. This would cause irreparable damage to thin-plated car batteries. Why?

Although lead sulfate crystals that form on the plates of a battery during deep discharge are removed when a battery recharges, some crystals fall off before recharge occurs. The thin plates of a car battery would therefore be quickly whittled away to nothing. After twenty or so deep discharges, the batteries would be ruined — no longer able to accept a charge. Being unable to accept a charge, they would have nothing to give back.

Renewable energy systems require deep-discharge lead-acid batteries with thick lead plates. The thickness of the lead plates allows them to withstand multiple deep discharges — so long as they're recharged soon afterwards. Even though the plates lose lead over time, they are so thick that the small losses are insignificant. As a result, a deep-cycle battery can be deeply discharged hundreds, sometimes a few thousand times, over its lifetime (see sidebar).

So, let there be no lingering doubts in your mind: car and truck batteries are no match for

A Word of Warning

Although batteries designed for deep discharge are considerably more robust than car and truck batteries, they are not invincible. If a battery is discharged too deeply — more than 80 percent of its stored capacity — it can be permanently damaged. Fortunately, controls in a well-engineered renewable energy system prevent this from occurring.

the powerful lead-acid batteries designed for the deep cycling that typically occurs in renewable energy systems. Also, for optimum long-term performance, remember that batteries need to be recharged promptly after deep discharging takes place. Don't ever forget this! Deep-cycle flooded lead-acid batteries like L16s that are promptly recharged after deep discharging, routinely equalized, and housed in a warm location, could (if treated well) last for seven to ten years, maybe even longer. Fail this routine maintenance and you'd better get your check book out.

Can I Use a Forklift, Golf Cart or Marine Battery?

Forklifts require high-capacity, deep-discharge batteries. These leviathans are designed for a fairly long life and operate under fairly demanding conditions. They can withstand 1,000 to 2,000 deep discharges — more than many other deep-cycle batteries used in battery-based renewable energy systems — and therefore are well suited for a renewable energy system.

Although forklift batteries function very well in a renewable energy system, they are rather heavy and expensive. If you can acquire them at a decent price, you may want to use them, if the size of your system warrants this large a battery.

Golf cart batteries may also work. Like forklift batteries, golf cart batteries are designed for deep discharge. They also typically cost a lot less than other heavier duty deep-cycle batteries. While the lower cost may be appealing, golf cart batteries don't last as long as the alter-

natives. They may last only five to seven years, if well cared for. Shorter lifespan means more frequent replacement. More frequent replacement means higher long-term costs and more hassle.

That said, Ian notes that golf cart batteries are an economical option and do work well in some renewable energy systems. Although they won't last as long as the heavier-duty deep-cycle batteries like L-16s, they are a good value. In fact, some renewable energy system installers swear by them, and use them in many of their systems. They are also often a good choice for first-time users, who are more likely to abuse their batteries and cause their early demise. It's better to learn on a cheaper battery.

What about marine batteries, that is, boat batteries? Although marine batteries are advertised as deep-cycle or "marine deep-cycle" batteries, they are really a compromise between a car starting battery and a deep-cycle battery. Their thinner plates just aren't up to the task of a renewable energy system. They will not last as long in deep-cycle service as true deep-cycle batteries.

What about Used Batteries?

Another option for cost-conscious home and business owners is a used battery. Although used batteries can often be purchased for pennies on the dollar, and may seem like a bargain, for the most part they're not worth it.

One reason is that many used batteries are discarded because they've failed or have experienced a serious decline in function. Another reason is that you have no idea how well — or more likely how poorly — they've been treated. How old are they? Have they been deeply discharged many times? Have they been left in a state of deep discharge for long periods? Have they been filled with tap water rather than distilled water? Have the plates been exposed to air due to poor maintenance? Although there are exceptions, most people who've purchased used flooded lead-acid batteries have been disappointed.

Don't Skimp on Batteries

When shopping for batteries for a renewable energy system, we recommend that you buy high quality deep-cycle batteries. Although you might be able to save some money by purchasing cheaper alternatives, including used batteries, frequent replacement is a pain in the neck. Batteries are heavy and it takes quite a lot of time to disconnect old batteries and rewire new ones. Bottom line: the longer a battery will last — because it's the right battery for the job and it's well made and well cared for — the better!

Sealed Batteries — A Misnomer?

The truth be known, sealed batteries are not totally sealed. Each battery contains a pressure release valve that allows gases and fluid to escape if a battery is accidentally overcharged. The valve keeps the battery from exploding. Once the valve has blown, though, the battery may need to be retired.

Sealed Batteries

While most off-grid renewable energy systems rely on flooded lead-acid batteries, grid-connected systems with battery backup often incorporate another type of lead-acid battery, sealed lead-acid batteries. Also known as captive electrolyte batteries, sealed batteries lack filler caps. They are filled with electrolyte at the factory, charged, and then permanently sealed. Because of this, captive electrolyte batteries are easy to handle and ship without fear of spillage. They won't even leak if the battery casing is cracked open, and they can be installed in any orientation — even on their sides. But most important, they never need to be watered.

Currently, two types of sealed batteries are available: absorbed glass mat (AGM) batteries and gel cell batteries. In absorbed glass mat batteries, thin absorbent fiberglass mats are placed between the lead plates. The mat consists of a network of tiny pores that immobilize the battery acid. The mat also creates tiny pockets that capture hydrogen and oxygen gases given off by the lead plates during charging. Unlike a flooded lead-acid battery, the gases can't escape. Instead, they recombine in the pockets, reforming water. That's why sealed AGM batteries never need watering.

In gel batteries, the sulfuric acid electrolyte is converted to a substance much like hardened jello by the addition of a small amount of silica gel. The gel-like substance fills the spaces between the lead plates.

Sealed batteries are often referred to as "maintenance-free" because fluid levels never need to be checked and because the batteries never need to be filled with water. They also never need to be (and should not be!) equalized, a process discussed later in this chapter. Eliminating routine maintenance saves a lot of time and energy. It makes sealed batteries a good choice for off-grid systems in remote locations where routine maintenance is problematic — for example, rarely occupied backwoods cabins or remote cottages. Sealed batteries are also used on sailboats and RVs where the rocking motion would spill the sulfuric acid of flooded lead-acid batteries.

Sealed batteries offer several additional advantages over flooded lead-acid batteries. One advantage is that they charge faster. Sealed batteries also release no explosive gases, so there's no need to vent battery rooms or battery boxes where they're stored. In addition, sealed batteries are much more tolerant of low temperatures. They can even handle occasional freezing, although this is never recommended. Sealed batteries also discharge more slowly than flooded lead-acid batteries when not in use. (All batteries self-discharge when not in use.)

Despite this impressive list of advantages, sealed batteries are much more expensive than flooded lead-acid batteries. They also store less electricity and have a shorter lifespan than flooded lead-acid batteries. In addition, they can't be rejuvenated like a standard lead-acid battery if left in a state of deep discharge for extended periods. During such times, lead sulfate crystals on the plates begin to grow. Over time, small crystals enlarge. Large crystals reduce battery performance by reducing its ability to store electricity. Batteries then take progressively less charge and have less to give back. Over time, entire cells may die, substantially reducing a battery bank's storage capacity.

Large crystals on the plates of flooded lead-acid batteries can be removed by a controlled overcharge, a procedure known as equalization. Although equalization is safe in unsealed flooded lead-acid batteries, it results in pressure buildup inside a sealed battery. Pressure is vented through the pressure release valve on the sealed battery, causing electrolyte loss that could destroy or seriously decrease the storage capacity of the battery. Finally, sealed batteries must be charged at a lower voltage setting than flooded lead-acid batteries.

So, while maintenance-free batteries may seem like a good idea, they are not suitable for many applications.

Wiring a Battery Bank

Batteries are wired by installers to produce a specific voltage and amp-hour storage capacity. Small renewable energy systems — for example, those used to power RVs, boats and cabins — are typically wired to produce 12-volt electricity. Many of these applications run entirely off 12-volt DC electricity. Systems in off-grid homes and businesses are typically wired to produce 24- or 48-volt DC electricity. The low-voltage DC electricity, however, is converted to AC electricity by the inverter, which also boosts the voltage to 120- and 240-volts, commonly used in homes and businesses.

Wiring a battery bank is beyond the scope of this book, but an understanding of some of the basics can be useful. One of them is that batteries can be wired in series or parallel. Wiring a battery bank in series means that the positive pole of one battery is connected to the negative pole of the next one, as shown in Figure 7.4. Wiring batteries in series increases the voltage. As an example, four 6-volt lead-acid batteries wired in series produce 24-volt DC electricity. Eight 6-volt batteries wired in series produces 48-volt DC electricity (Figure 7.4). A group of batteries wired in series forms a string.

While wiring in series boosts the voltage, wiring in parallel increases the amp-hour storage

Fig. 7.4: *In the upper diagram, four 6-volt batteries are wired in series to produce 24 volts. In the lower diagram, eight 6-volt batteries are wired in series to produce 48-volts.*

Fig. 7.5: *Each string of 6-volt batteries is wired in series, yielding 48 volts. By wiring in parallel (connecting positive and negative terminals of the end batteries in each series), you double the amp-hour storage capacity.*

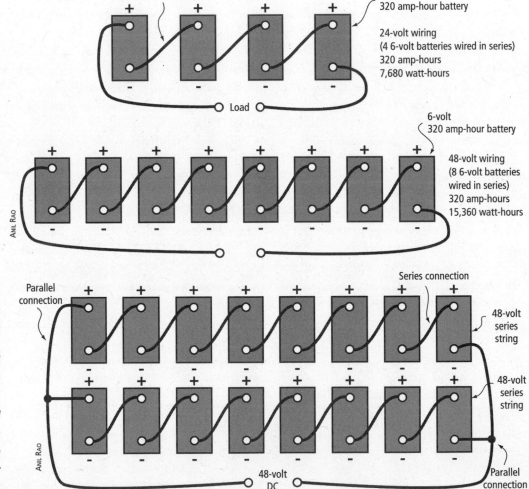

capacity of a battery bank — how many amp-hours of electricity a battery bank can store. To boost the amp-hour capacity of a renewable energy system, installers typically include two or more strings of batteries in parallel, as shown in Figure 7.5. As you can see from the diagram, each string is wired in series, but the strings are then connected in parallel.

As shown in Figure 7.5, in parallel wiring the positive ends of each string are connected, as are the negative ends. Battery banks wired in this manner have the same voltage as the individual strings, but a higher amp-hour rating.

Sizing a Battery Bank

Properly sizing a battery bank is key to designing a reliable off-grid wind or hybrid wind/solar electric system. The principal goal when sizing a battery bank is to install a sufficient number of batteries to carry your household or business through periods when the wind or wind and sun are not available, in the case of a hybrid wind and solar system.

The easiest way to size a battery bank is to begin by calculating how much electricity you use in a day. Because electrical consumption varies from month to month, your best bet is to determine monthly consumption and then determine the daily rate during the month of greatest consumption, a topic discussed in Chapter 4.

Once you determine how many kilowatt-hours of electricity you will consume per day during your most energy-intensive month, you must make an educated guess as to the number

Wiring Advice

For optimum performance, it is important to reduce the length of battery cable when wiring a battery bank. Place batteries close to the charge controller and inverter and keep battery cables as short as possible.

of windless days that occur in that time. That's how long your renewable energy system will be sidelined by a lack of wind. In most cases, renewable energy designers plan for three days of no wind. Battery banks are then sized to meet one's electrical needs for three days. Longer reserve periods — five days or more — may be required in some areas.

As noted in Chapter 3, backup fossil-fuel generators are often included in off-grid systems. Backup generators can reduce the need for a large battery bank. Then the main question becomes, "How frequently do I want to operate my generator?"

Let's look at an example to see how this works. Let's suppose that you or your business consume four kilowatt-hours (4,000 watt-hours) of electricity a day on average during the most energy-intensive month. If you need five days of battery backup, you will need 20 kilowatt-hours of storage capacity (5 days x 4 kilowatt-hours).

However, it is best not to discharge batteries below the 50 percent mark as deeper discharging rates reduce battery life. If you set 50 percent as your discharge goal, then you'll need 40 kilowatt-hours of storage capacity.

To determine how many batteries you will need, you will need to determine the amp-hour storage capacity of the batteries you are thinking about installing. An amp-hour refers to electricity flowing out of a battery at a rate of one amp for one full hour. A battery, for instance, might be rated at 300 amp-hours. How do you know how many amp-hours of storage capacity you need when all you know is that you need 40 kilowatt-hours of storage capacity?

To convert kilowatt-hours to amp hours, simply divide the kilowatt-hours by the voltage of the system. (Remember the equation watts = amps x volts. To solve for amp-hours, divide watt-hours by volts). If you are installing a 24-volt system, 40,000 watt-hours divided by 24 volts results in 1,666 amp-hours. That's the storage capacity required for three days.

The next task is to see how many batteries you will need to store this much electricity. You do that by checking the manufacturer's specs on their batteries. The Trojan L-16Hs Dan uses in his hybrid solar electric/wind system are six-volt batteries that have a rated capacity of 420 amp-hours at six volts. To determine how many batteries he'd need in this example, you'd think that all you would need to do is divide 1,666 amp-hours of storage capacity by 420 amp-hours of battery storage. The result would be four batteries.

Well, not quite.

There's a little more to it. If you did this, you'd be sadly disappointed.

If you recall from previous discussions, wiring four six-volt batteries in series increases the voltage to 24 volts, but does not change the amp-hour storage capacity. That is to say, a string of four 6-volt batteries, each with a 420 amp-hour capacity, will provide 24-volt DC electricity, but will yield only 420 amp-hours of electricity. To create 1,666 amp-hours of storage, you'd need four strings, each one of which contained four 6-volt batteries. Each string would yield 420 amp-hours of electricity. Combined in parallel, the four strings would provide the 1,600 plus amp-hours required to meet your needs. So to meet your needs, you would need 16 batteries.

Battery Storage: Watt-Hours vs. Amp-Hours?

Although battery capacity is typically measured in amp-hours, this rating system can cause some confusion. A much more useful way to rate batteries would be by watt-hours (and kilowatt-hours). As Ian notes, we measure loads in watt-hours, energy production in watt-hours, and utility energy purchased and sold in watt-hours. Why not batteries, too?

Amp-hours may be a trifle more technically accurate when it comes to batteries (for reasons we won't go into here). At the level of accuracy of home systems, however, it's not worth the added confusion to use amp-hours. Measuring load, energy production, and battery storage all in one unit would make all calculations easier.

To calculate watt-hours, multiply the individual battery or battery string amp-hours by the individual battery or string voltage. If, for example, you have one six-volt battery with a 200 amp-hour rated capacity, you would multiply the voltage by the amp-hour capacity. This battery therefore stores 1,200 watt-hours of electricity, or 1.2 kWh. If we wire four six-volt, 200-amp-hour batteries in series, the string voltage is 24. The amp-hour storage capacity, however, remains 200. To determine the watt-hours stored in the battery, you multiply 6 volts x 4 batteries x 200 amp-hours, which equals 4,800 watt-hours or 4kWh or we could multiply 24 volts by 200 amp-hours to come up with the same result.

If batteries were rated in watt-hours, we'd lose most of the slightly confusing math, and remove the opportunities for error and misunderstanding. The batteries above would be rated at one kilowatt-hour each, and no matter how we wire them (series, parallel, or series-parallel), we'd just add up the kilowatt-hour total, to determine their storage capacity.

As you can imagine, storage can become expensive. In the example we just used, 16 batteries could cost $4,500 (not including installation costs).

Because of the cost and because batteries require care and periodic maintenance and take up a lot of room, many homeowners install hybrid systems — typically, wind and PV systems. This allows homeowners to reduce the size of their battery banks. Most homeowners throw in a gen-set — a gasoline, diesel, propane or natural gas generator — to provide additional backup power.

Hybrid systems and gen-sets reduce battery storage. Although solar arrays and gen-sets aren't cheap, solar modules won't need to be replaced after ten years like a battery would. (Gen-sets may need replacement if used heavily.) PV modules will very likely outlast four or five battery banks, so the additional generating capacity may be well worth your investment in the long run. In addition, PV systems require very little maintenance. Ian points out, however, "Though I love the theory of investing in renewable energy capacity instead of lead, there is a limit to its potential." In cloudy regions, PVs may not produce enough electricity to make a huge difference during winter months. In such instances, batteries and gen-sets may be a preferred option. How do you know what to do?

Fortunately, the National Renewable Energy Laboratory offers online assistance. The program, called HOMER, is designed to optimize hybrid power systems. It can be found at analysis.nrel.gov/homer. "This is a great tool for optimizing the size of various components in a hybrid power system," notes Jim Green, although it is a bit difficult for the average homeowner. "It typically shows that adding a backup generator will reduce system cost and that optimum battery bank size may be on the order of 8 to 12 hours of storage, much less than 3 days, which is typically used." He adds, "Of course, the optimization point will change over time as the price of propane and natural gas go up. Even so, the generator run-time can be quite low."

Battery Maintenance and Safety

Now that you understand lead-acid batteries, your options, and a little about wiring and sizing a battery bank, it is time to turn to battery care and maintenance and safety. Battery care and maintenance are vital to the long-term success of battery-based renewable energy systems. Proper maintenance increases the service life of a battery. Because batteries are expensive, longer service life results in lower operating costs over the long haul. The longer your batteries last, the cheaper your electricity will be.

Keep Them Warm

Batteries may be the workhorses of renewable energy systems, but they're also sissies. They like to be kept warm and they like it best when they have full tummies. Cold dramatically reduces the amount of electricity they'll store and yield (Figure 7.6). Low temperatures slow down the chemical reactions in batteries, reducing electrical storage.

While low temperatures reduce battery efficiency, higher temperatures increase outgassing, the release of explosive hydrogen gas,

and water loss, which reduces battery fluid levels. Outgassing is not a trivial matter. Higher temperatures also lead to higher rates of self-discharge in old and new lead-acid batteries. (As a rule, older batteries lose charge faster than new batteries.)

For optimal function, then, batteries should be kept at around 75 to 80 degrees Fahrenheit. Batteries like to live at about the temperature we find comfortable. In this range, they'll accept and deliver tons more electricity. Guaranteed!

If you can't maintain batteries in this narrow temperature range, try to shoot for a range between 50 and 80 degrees Fahrenheit. Rarely should batteries fall below 40 degrees or exceed 100 degrees Fahrenheit.

Ideally, batteries should be stored in a separate battery room or a battery box inside a conditioned space to maintain the optimum temperature. We discourage people from housing their batteries inside their homes and recommend housing them in a heated space like a shop, utility room or garage, although Dan built a sealed battery closet in his utility room, which works well.

Whatever you do, don't store batteries in a cold garage, barn or shed. Besides delivering less electricity, they won't last long. In addition, if the batteries are in a low state of charge, they can freeze. Freezing may cause the cases of flooded lead-acid batteries to crack, leaking electrolyte and creating a dangerous mess in the battery room.

Batteries should not be stored on concrete floors. Cold floors cool them down and reduce their capacity. Always raise batteries off the floor.

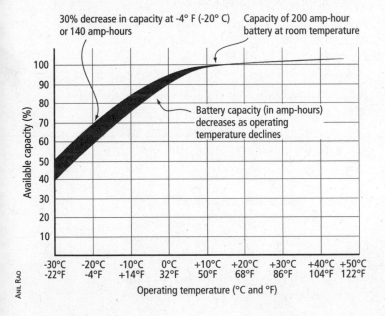

ANIL RAO

30% decrease in capacity at -4° F (-20° C) or 140 amp-hours

Capacity of 200 amp-hour battery at room temperature

Battery capacity (in amp-hours) decreases as operating temperature declines

Available capacity (%)

Operating temperature (°C and °F)

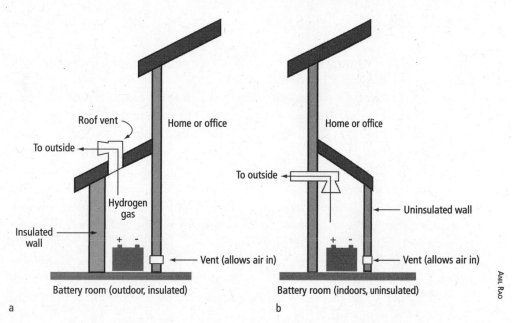

Roof vent

To outside

Home or office

Hydrogen gas

Insulated wall

Vent (allows air in)

Battery room (outdoor, insulated)

a

Home or office

To outside

Uninsulated wall

Vent (allows air in)

Battery room (indoors, uninsulated)

b

ANIL RAO

If you must store batteries outside your home, be sure to heat and cool the space. Heating and cooling a shed, garage or barn can be quite expensive and wasteful of costly fossil fuel energy. If you must heat a battery room, we strongly recommend retrofitting or building a storage facility that's passively heated and cooled — or perhaps installing a solar hot air collector. Insulate the building to the max and provide south-facing glass to keep it warm — but not too warm — in the winter. (You may want to read more on passive solar design in one of Dan's books, *The Homeowner's Guide to Renewable Energy* or *The Solar House*. The latter offers a more detailed account.)

Ventilate Your Batteries

To ensure safe operation, battery rooms should be ventilated to remove potentially explosive hydrogen gas released when batteries charge. Never place batteries in a room with a gas-burning appliance or an electrical device such as an inverter, even if the enclosures are vented. A tiny spark could ignite the hydrogen gas, causing an explosion. Also be sure that activities that require an open flame or that might produce sparks are never carried out in the vicinity of a battery bank.

As shown in Figure 7.7, proper ventilation requires small air openings that allow fresh air into the battery box or battery room near floor level. It also requires an air outlet near the top of the box or ceiling to vent hydrogen gas. Check with local building codes to be sure you comply with their requirements, if any. A small two-inch vent made from PVC or ABS pipe is all that's typically required to vent a battery room or battery box. Be sure that the vent system doesn't cool the battery room down in the winter, however.

Although most battery rooms are passively ventilated, they can also be actively vented or power vented. Power venting requires a small no-spark electric fan that exhausts hydrogen gas in the enclosure while batteries are charging. As nifty as this may seem, it's generally not necessary. Hydrogen is an extremely light gas and easily escapes if a room is properly vented. So unless the vent pipe is long and contorted, you probably don't need to power vent. If, however, batteries are located in a room used for other purposes, for example, a shop or garage, active venting may be helpful. This reduces the chance of explosions and also rids the room of smelly gases produced when batteries are charging. (To learn about power venting from one manufacturer, log on to zephyrvent.com.)

Battery Boxes

Large battery banks are often housed in separate, lockable and ventilated rooms. Rather than dedicating an entire room to batteries, however, many homeowners install batteries in sealed and ventilated battery boxes in rooms that serve other purposes. Battery boxes are typically built from plywood. Some individuals build battery boxes with solid wooden lids; others install clear plastic (polycarbonate) lids. Be sure to line the boxes with an acid-resistant liner to contain possible acid spills. Lids should be hinged and sloped to discourage people from storing items on top. Also, be sure to place battery boxes in a warm room. Raise the boxes off concrete floors. Be

sure to seal all battery boxes and vent them to the outside.

Batteries can also be stored in battery boxes made from off-the-shelf plastic tubs, althouh they will need to be modified to provide ventilation and should be raised off the floor.

Some individuals purchase battery storage boxes from commercial outlets. Radiant Solar Tech at radiantsolartech.com, for example, manufactures sealed plastic battery boxes with removable lids for renewable energy systems. This company will also custom build boxes for homeowners.

When shopping for battery boxes, note that many commercially available battery boxes are designed for sealed batteries, not the flooded lead-acid batteries. Before you buy a manufactured battery storage container, be sure that it will work with the batteries you are planning on installing. Containers for flooded lead-acid battery banks should provide sufficient clearance to view electrolyte levels and to fill cells when battery fluid levels drop and should also be vented.

Finally, batteries should be located as close to the inverter and other power conditioning equipment as possible. Doing so minimizes voltage drop and power losses and will also help reduce system costs.

Keep Kids Out

If young children are present in a home or business, battery rooms and battery boxes should be inaccessible and locked. This will prevent children from coming in contact with the batteries, and risking electrical or acid burns. Although electrocution is not a hazard at 12, 24 or 48 volts, dropping a tool or other metal object on the battery terminals could result in an electrical arc that can cause burns, or could result in an explosion of a battery case resulting in acid burns.

Avoiding Deep Discharge to Ensure Longer Battery Life

Keeping flooded lead-acid batteries warm and topped off with distilled water ensures a long life span and optimum long-term output. Longevity can also be ensured by keeping batteries as fully charged as possible. To understand this issue, we first explore a subject briefly mentioned earlier in the chapter: cell capacity — the amount of electricity a battery can store.

"Cell capacity is the total amount of electricity that can be drawn from a fully charged battery until it is discharged to a specified battery voltage," writes Richard J. Komp, author of *Practical Photovoltaics: Electricity from Solar Cells*. Battery cell capacity is measured in amp-hours. A Trojan L16H battery can cycle approximately 420 amp-hours of electricity. But what does this mean?

A battery with a 420-amp-hour storage capacity suggests that it would theoretically deliver one ampere of electricity over a 420-hour-period or 420 amperes in one hour. In reality, that's not what happens. Battery capacity varies with the discharge rate — how fast a battery is discharged. As a rule, the faster a lead-acid battery is discharged, the less you'll get out of it. A lead-acid battery discharged over a ten-hour period, for example, will yield 100 percent of its rated capacity.[1] Discharge that battery in an hour and a half, and it will deliver only 75 percent of its rated capacity. Figure 7.8 illustrates the energy removed from batteries at various discharge rates.

Most batteries are rated at a specific discharge rate, usually 20 hours (which is what the 420 AH capacity in the example above is based on). What does battery capacity have to do with life span?

Like so many things in life, lead-acid batteries last longer the less you use them. That is to say, the fewer times a battery is deep discharged, the longer it will last. As illustrated in Figure 7.9, a lead-acid battery that's frequently discharged to 50 percent will last for slightly more than 600 cycles, if recharged after each deep discharge. If discharged no more than 25 percent of its rated capacity — and recharged after each deep discharge — the battery

Fig. 7.8: *Amp-hours vs. Rate of Discharge. The rate of discharge dramatically influences the number of amp-hours a battery will yield, as shown here. The faster a battery is discharged the lower its capacity. The slower a battery is discharged, the higher its output, measured in amp-hours.*

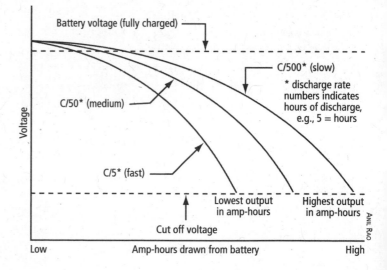

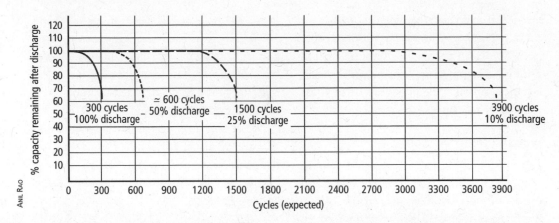

ANIL RAO

Fig. 7.9: *Battery Life vs Deep Discharge Shallow discharging prolongs battery life as explained in the text, while deep discharging reduces battery life. Note the difference in the number of cycles a battery can undergo at routine 50 percent discharge vs. 10 percent discharge.*

should last about 1,500 cycles. If the battery is discharged only 10 percent of its capacity, it will last for 3,600 cycles.

But wait, you say, "I'm buying deep-cycle batteries that can handle deep discharges."

Yes, but frankly, the less often you do so, the longer your batteries will last. As many renewable energy folks will tell you, shallow cycling makes batteries last longer.

As Ian points, out, however, this topic (like so many issues in life) is complicated. While deep discharging reduces the lifespan of a battery, renewable energy users are more concerned with the cost of the battery per watt-hour cycled. In other words, what we want from batteries is not simply to last a long time but to cycle a lot of energy.

Studying the cost of a battery per cycled kilowatt-hour, says Ian, provides a better measure than total years of service. Theoretically, you'll get the most bang for your buck by cycling in the 40 percent to 60 percent deep discharge range. This goes against the "shallow cycling makes batteries last longer" idea.

In off-grid renewable energy systems, it is also important to never leave batteries at a low state of charge for a long time. Even more important than avoiding deep discharges then is ensuring prompt full recharges. Achieving these goals is easier said than done, however. If your system is small and you don't pay much attention to electrical use, you'll very likely overshoot the 40 to 60 percent mark time and time again. If you carefully monitor your electrical usage, and charge batteries soon after deep discharges, you are more likely to achieve your goals.

One way of reducing deep discharge is to conserve energy and use electricity as efficiently as possible. Conserving energy means not leaving lights and electronic devices running when they're not in use — all the stuff your parents told you when you were a kid. It means ridding your home or business of phantom loads. Energy efficiency also means installing energy-efficient lighting, appliances, electronics, and so on — the ideas many energy conservation experts have been suggesting for decades.

Conserving energy and making your home or business energy-efficient is only half the battle, however. You may also have to adjust electrical use according to the state of charge of your batteries. In other words, you have to cut back on electrical usage when batteries are more deeply discharged and shift demand for electricity to times when the batteries are more fully charged. You may, for instance, run your washing machine and microwave when the wind's blowing and your batteries are full, but hold off when the wind's stopped and batteries are running low — unless you want to run a backup generator. How do you monitor the state of charge in your batteries?

After living with solar electricity and wind energy since 1996, Dan does this by watching the weather. If he has experienced a couple of cloudy, windless days, he knows that he has to curtail his electrical consumption. He may cook on the gas-powered range, as opposed to the microwave. He may hold off on laundry until the sun comes out or the wind blows again.

Dan monitors his system by checking the battery voltage on the meter in his power control center at the end of the day (when the sun's not shining and the wind's not blowing and there's little, if any, electricity being drawn from the battery bank). The voltage reading gives him a clue about the state of charge of his batteries.

Voltage is a pretty crude way of monitoring battery capacity, but it is better than nothing, if you know how to interpret it. If the batteries are truly "at rest," low voltage readings mean the batteries are in a low state of charge and higher readings indicate that the batteries are full. Just remember, batteries' voltage and state of charge are not directly related unless the batteries are at rest. That means they are not being charged or discharged and have not been recently charged or discharged (they need a little time without either for the voltage to settle). For those who are interested in learning more, battery manufacturers provide charts that correlate state of charge with charge/discharge rate and voltage.

A more precise way to monitor the state of batteries is through the use of a digital amp-hour or watt-hour meter. One popular meter, the TriMetric, keeps track of the number of amp-hours of electricity stored in a battery bank each day. (That is, the number of amp-hours produced by the renewable energy system and stored in the batteries.) It also indicates the number of amp-hours drawn from the batteries. In addition, this meter keeps track of the total amount stored in a battery bank at any one time, that is, how full the batteries are. To make your life easier, the TriMetric displays a "fuel gauge," a reading that shows the state of charge as a percentage. Many laptop computers have similar state-of-battery-charge indicators for their batteries.

As noted above, homeowners can use this battery information to adjust their activities. If batteries are approaching the 40 to 60 percent discharge mark, they may hold off on activities that consume lots of electricity. Or, they may elect to run a backup generator to charge the batteries.

Another way to minimize deep discharging is to oversize a wind system. One way to do this is to install a larger wind turbine on the tower — one with a greater annual energy output than your site analysis suggests. However, the most economical way of boosting energy production in a wind energy system is to increase tower height. Another option is to install a hybrid system — a wind machine and a solar electric array. This is especially useful if the wind does not blow year round, which is typical in most locations. Hybrid systems ensure a steady year-round supply of electricity.

A backup generator can also be installed as a third source of electricity. Backup generators can be fired up when batteries are running low. They can also be wired directly to the inverter so they turn on automatically when the battery voltage drops to a predetermined level. Bear in mind that not all inverters and not all generators operate automatically, so be sure your generator is compatible with this application and that your inverter contains the proper circuitry.

Watering and Cleaning Batteries

The lead plates in batteries are immersed in a 30 percent solution of sulfuric acid. As noted earlier, sulfuric acid participates in reversible chemical reactions that store and release electricity. Because these reactions are reversible, you'd think that battery fluid levels would remain constant over the long haul. Unfortunately, that's not the case. Battery fluid levels decrease over time because water is also lost when batteries are charged.

Electricity flowing into batteries splits water molecules in the electrolyte into their component parts — hydrogen and oxygen. This process is known as electrolysis. (Electrolysis is the source of the potentially explosive mixture of hydrogen and oxygen gas that makes battery room venting necessary.) Hydrogen and oxygen produced during electrolysis are both gases. These gases can escape through the vents in the battery caps in flooded lead-acid batteries, lowering fluid levels. In addition, water can evaporate through the vents, and a mist of sulfuric acid can escape through the vents during charging, also depleting fluid levels in batteries.

All of these sources of water loss add up over time and can run a battery dry. When the plates are exposed to air, they quickly begin to corrode. When this happens, a battery's life is pretty well over. Dan's tried all kinds of tricks to bring batteries that have fallen victim to low-electrolyte levels back to life, but to no avail. Making matters worse, when one cell in a battery goes bad, the entire battery is shot. One bad battery renders a whole battery bank bad. If the battery bank is more than a year to a year and a half old, you can't simply replace a bad battery with a new one. You have to replace all your batteries.

To prevent this potentially costly occurrence, check battery fluid levels regularly. Many experts recommend checking batteries on a monthly basis. Others recommend checking batteries every three months. Dan has found that a two- to three-month checkup works well in his system.

Be careful not to get lulled into complacency after installing new batteries, however. The first year of a battery's existence is like the first year or two of marriage. Peace and happiness prevail. Brand-new batteries operate very smoothly for a year or so with very little water loss. As time passes, however, batteries become a lot more demanding and require more frequent inspection and much more care.

To check fluid levels, unscrew the caps and peer into each cell when the batteries are not charging. Use a flashlight, if necessary — never a flame from a cigarette lighter! Battery acid should cover the lead plates at all times — at a bare minimum a quarter of an inch above the plates. As a rule, it is best to fill batteries just below the bottom of the fill well — the opening in the battery casing into which the battery cap is screwed.

When filling a battery be sure to only add distilled or deionized water. Distilling and deionizing are different processes that produce similar quality pure water. Never use tap water. It may contain minerals or chemicals that contaminate the battery fluid, reducing a battery's life span.

Batteries should be fairly well charged before topping them off with distilled water. Don't fill batteries, and then charge them, for example, with a backup generator. Overfilling a battery could result in battery acid bubbling out of the cells when the batteries are charged — either by the wind turbine or a backup generator. If fluid level is extremely low, add a little distilled water before charging them.

Electrolyte loss in overly filled batteries not only reduces battery acid levels, it also deposits acid on the surface of batteries. When it dries, the acid forms a white coating. This not only looks messy, it can conduct electricity, slowly draining the batteries. Battery acid also corrodes metal — electrical connections, battery terminals and battery cables. In addition, the loss of battery acid can result in a dilution of the electrolyte within the battery cells because the addition of distilled water to compensate for lost fluid will dilute the acid remaining in the cells.

Battery acid bubbling out of batteries needs to be cleaned promptly. Use distilled water and paper towels or a clean rag. Some sources recommend using a solution of baking soda (sodium bicarbonate) to neutralize battery acid that bubbles up onto the top of batteries. After carefully rinsing batteries with sodium bicarbonate, they say, batteries should be wiped down with distilled water. We discourage people from cleaning batteries with baking soda because it could drip into the cells of a battery, neutralizing the acid and reducing battery capacity. Distilled water is sufficient to clean the surface of batteries. When cleaning batteries, be sure to wear gloves, protective eyewear and a long-sleeved shirt — one you don't care about. If you get acid on your skin, wash it off immediately with soap and water.

Although we don't recommend using baking soda to clean batteries, it is not a bad idea to keep a few boxes on hand just in case there's an acid spill in the battery room. Acid spills are extremely rare but may occur if a battery case cracks. You should also install a fire extinguisher for fire safety. Check with the battery manufacturer for the type of extinguisher they recommend.

When filling batteries, be sure to take off watches, rings and other jewelry, especially loose-fitting jewelry. Metal jewelry will conduct electricity if it contacts both terminals of a battery. Such an event will leave your jewelry in a puddle of metal — along with some of your flesh. One 6 or 12 volt cell can produce more than 8,000 amps if the positive and negative terminals of a battery are connected. In addition, sparks could ignite hydrogen and oxygen gas in the vicinity, causing an explosion. Shorting out a battery can also crack the case, releasing battery acid.

Also be careful with tools when working on batteries — for example, tightening cable connections. A metal tool that makes a connection between oppositely charged terminals on a battery may be instantaneously welded in place. The tool will become red hot and could also ignite hydrogen gas, causing an explosion. It will also very likely ruin the battery. Wrap hand tools used for battery maintenance in electrical tape so that only one inch of metal is exposed on one end; that way it can't make an electrical connection. Or buy insulated tools to prevent this from happening.

Even if you are careful not to overfill your batteries, you will very likely need to clean the battery posts every year or two. As noted earlier, under normal operation batteries release a fine corrosive mist (containing sulfuric acid) when charging. The acid can react with metal posts, battery cables and nuts that attach the battery cables to the posts, causing corrosion.

To clean the posts, use a small wire brush, perhaps in conjunction with a spray-on battery cleaner purchased at a local hardware store. To reduce maintenance, coat battery posts with Vaseline or a battery protector/sealer, available at hardware and auto supply stores. This protects the posts as well as the nuts that secure the battery cables on the posts.

Equalization

Recharging batteries and maintaining proper fluid levels ensures optimum function over the long haul. But that's not all you can do to extend your batteries' service life. To get the most out of batteries, you also need to periodically equalize them. Equalization, mentioned earlier, is a controlled overcharge of batteries.

Why Equalize?

Periodic equalization is performed for three reasons. The first is to drive lead sulfate crystals off the lead plates, preventing the formation of larger crystals. As noted earlier, large crystals reduce battery capacity, and are difficult to remove. As a result, they can permanently damage the lead plates.

Batteries must also be periodically equalized to stir the electrolyte. Sulfuric acid tends to settle near the bottom of the cells in flooded lead-acid batteries. During equalization, hydrogen and oxygen gases released by the breakdown of water (electroylysis) create bubbles that cause the battery to "boil." (This phenomenon is called boiling because the rapid formation of bubbles resembles boiling water. It is not the same as boiling water, however, as it is not caused by high temperatures.) These bubbles, in turn, mix the fluid so that the concentration of acid is equalized throughout each cell of each battery, ensuring better function.

Equalization also helps bring all of the cells in a battery bank to the same voltage. That's important, indeed vital, to a battery bank because some cells may sulfate more than others. As a result, their voltage may be lower. A single low-voltage cell in one battery reduces the voltage of the entire string. In many ways, then, a battery bank is like a camel train. It travels at the speed of the slowest camel.

If equalization restores batteries, why won't a battery remain functional for an eternity? Although equalization removes lead sulfate from plates, restoring their function, some lead dislodges — or flakes off — during equalization, settling to the bottom of the batteries. As a result, batteries lose lead over time and never regain their full capacity.

How to Equalize

Equalizing batteries is a simple process. In those systems with gen-set for backup, the owner simply sets the inverter to the equalization mode and then cranks up the generator. The inverter controls the process from that point on. In wind/PV hybrid systems, the operator sets the wind generator controller to the equalize setting during a storm or period of high wind. The controller takes over from there.

During equalization, the number of amps fed into a battery bank (the charge amperage) is kept relatively constant and high. During this process, the battery voltage is allowed to rise to higher than normal. (Normally the charge controller shuts off the flow of electricity to batteries once a certain voltage is reached. As noted in Chapter 3, this is called overcharge

protection.) Over time, the voltage of all of the cells of all the batteries in a battery bank rise to the same level, equalizing them. This process usually takes four to six hours. Once the batteries are equalized, the process is terminated automatically.

During equalization, voltage may rise quite high. For a 12-volt system, the voltage may rise as high as 15 to 16 volts. In a 24-volt system, it may rise to 30 to 32 volts. Ian reminds us that loads sensitive to high voltage should be disconnected during equalization, although in modern all-AC home systems, the inverter will protect the system because it contains a high-voltage shut-down switch.

How often should batteries be equalized?

The answer to this question depends on whom you talk to and how hard you work your batteries. Some installers recommend equalization every three months. If your batteries are frequently deep discharged, however, you may want to equalize more frequently. If batteries are rarely deep discharged, they'll need less frequent equalization. For example, batteries that are rarely discharged below 50 percent may only need to be equalized every six months.

The frequency of equalization also depends on your normal charge set points. Some homeowners run their battery banks "hot" — at a higher voltage setting than the standard. So the batteries need less equalization, but more watering.

Rather than second guess your batteries' needs for equalization, it is wise to check the voltage of each battery, using a volt meter, every month or two. If you notice that the voltage of

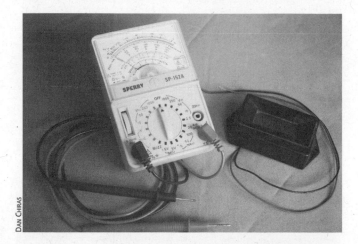

DAN CHIRAS

Fig. 7.10: *Meters like these are used to check the voltage of individual batteries to troubleshoot a system. They help a system operator determine whether equalization is required and if individual batteries have gone bad.*

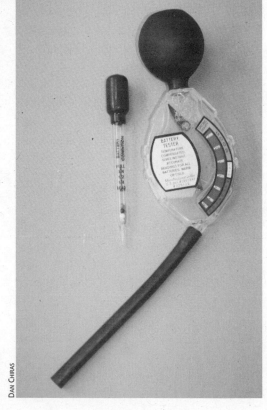

Fig. 7.11: *Hydrometers measure specific gravity. Low specific gravity indicates that the battery needs recharging.*

DAN CHIRAS

one or two batteries is substantially lower than others, equalize the battery bank. Checking voltage requires a small voltmeter like the one shown in Figure 7.10. Dan uses a standard portable electrical meter. Digital volt meters work best — they're easier to read.

Another way to test batteries is to measure the specific gravity of the battery acid using a hydrometer (Figure 7.11). Specific gravity is a measure of the density of a fluid. Density is related to the concentration of battery acid — the higher the concentration, the higher the specific gravity. If significant differences in the specific gravity of the battery acid are detected in the cells of a battery bank, it is time to equalize.

Checking the voltage of the batteries and the specific gravity of the cells may involve more work than you'd like and may not be necessary if you pay attention to weather and battery voltage or adhere to a periodic equalization regime. Dan rarely equalizes his batteries during Colorado's sunny summer months, as his batteries are often full to overflowing with the electricity produced by his solar electric array. In the winter, however, he equalizes every two to three months. When batteries run low, he may run the generator for an hour or two to bring the charge up to prevent deep cycling. This is not an equalization, just an attempt to recharge the batteries more often in cloudy weather.

If at all possible, Mick recommends using a wind turbine to equalize the batteries of an off-grid system. You'd be amazed at how well this works, as was Johnny Weiss of Solar Energy International in sunny Colorado. Johnny was astonished when he observed a newly installed wind turbine charge a battery bank in a PV/wind hybrid system. The wind turbine, which was installed in a workshop taught by Mick, was added because the client wanted to wean herself from a gasoline-powered gen-set. After living with the wind/PV hybrid system for a year, the owner sold the gen-set as she no longer needed it. Ian points out that conditions have to be right for this strategy to work — that is, you need to have good solar *and* wind resources, and may need to be willing to curtail usage when both resources are scarce.

As a final note on the topic, be sure only to equalize flooded lead-acid batteries. A sealed battery, either gel cell batteries or absorbed glass matt sealed batteries, cannot be equalized! If you try to, you'll ruin them. Also, don't equalize batteries fitted with Hydrocaps, discussed next. Hydrocaps must be removed prior to equalization or you will ruin them.

Reducing Battery Maintenance

Battery maintenance should take no more than 30 minutes a month. (It takes about ten minutes to check the cells in a dozen batteries but may take twenty minutes to add distilled water to each cell if battery fluid levels are low.)

If this is more work than you're interested in taking on, you'll be delighted to learn that there are a number of ways that battery maintenance can be reduced. One way is to install sealed batteries, although they're best suited for grid-connected systems with battery backup.

Another way to reduce battery maintenance is to replace factory battery caps with Hydrocaps, shown in Figure 7.12. Hydrocaps

JOE SCHWARTZ

Fig. 7.12: *Hydrocaps. These simple devices help reduce battery watering by reducing water losses.*

capture much of the hydrogen and oxygen gases released by batteries when charging under normal operation. They contain a small chamber filled with tiny beads coated with a platinum catalyst. Hydrogen and oxygen that have escaped from the cells react in the presence of this catalyst to form water, which drips back into the batteries. Because of this, Hydrocaps reduce water losses by around 90 percent and thus greatly reduce battery filling.

Another option is the Water Miser cap. Water Miser caps are molded plastic flip-top vent caps. Plastic pellets inside these caps capture moisture and acid mist escaping from batteries' fluid. This reduces sulfuric acid fumes in the battery room and prevents terminal corrosion. Unlike Hydrocaps, they don't capture hydrogen and oxygen formed from the breakdown of water during charging. Water Miser caps only reduce water loss by around

30 to 75 percent, but they can be left in place during equalization.

Of the two, we prefer Hydrocaps. They cost a bit more and must also be removed when equalizing batteries, but they reduce water losses more significantly.

Yet another way to reduce maintenance is to install a battery filling system, shown in Figure 7.13. Battery filling systems consist of a series of plastic tubes that are connected to special battery caps on each cell. The battery caps are fitted with float valves. In fully automatic systems, the plastic tubing is connected to a central reservoir containing distilled water. Valves in the battery caps open when the electrolyte level in cells drops. Distilled water flows by gravity from the central reservoir. When the cells are full, the float valves stop the flow of water.

In manually operated systems, the plastic tubing is connected to a tube fitted with a hand-operated pump. It is connected to a distilled water supply, like a one-gallon jug of distilled water. A quick coupler allows the tubing to be connected to the pump and the distilled water supply. Pressure buildup tells the operator when all the cells have been filled to the proper level. The entire process takes less than five minutes.

Figure 7.13: Battery Filling System. Distilled water can be fed automatically to battery cells or manually pumped into them through plastic tubing. Both approaches save a lot of time and energy and help to keep battery fluid levels topped off to ensure battery longevity.

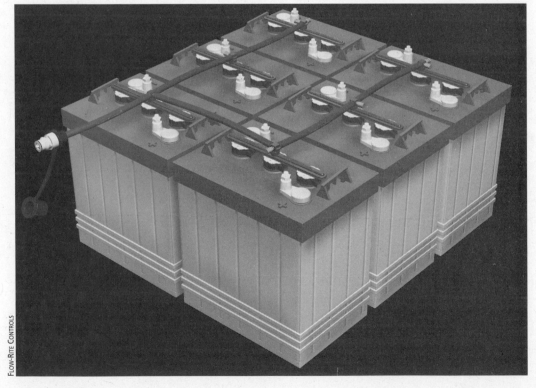

FLOW-RITE CONTROLS

Dan and Ian use manually operated Qwik-Fill battery watering systems manufactured by Flow-Rite Controls in Grand Rapids, Michigan, and sold online through Jan Watercraft Products. They've found that this system works extremely well after many years of service and has turned battery maintenance from a chore to a pleasure.

Although battery-filling systems work well, they're costly. Dan spent about $300 for his 12 batteries. Although that's a bit pricey, he, like Ian, knows that the systems quickly pay for themselves in reduced maintenance time and ease of operation. Both of us find that the convenience of quick battery watering overcomes the procrastination that leads to costly battery damage.

A simpler and cheaper alternative is a half-gallon battery filler bottle (Figure 7.14). It contains a spring-loaded valve in the spout (similar to the float valves in battery filling systems). The valve shuts off the flow of distilled water when the battery fluid level comes to within an inch of the top of the cell. You can order battery filler bottles from local battery suppliers or online from several suppliers for $10 to $20.

Living with Batteries

Batteries work hard for those of us who live off-grid. As you have seen, flooded lead-acid batteries need to be kept in a warm place, but not too warm. Enclosures for flooded lead-acid batteries need to be vented, too, and batteries need to be kept clean. They also need to be periodically filled with distilled or deionized water. You also need to monitor their state of

DAN CHIRAS

charge and either charge them periodically with a backup supply of power when they're being overworked or back off on electrical use. You don't want them to sit in a state of deep discharge for too long.

Generators in such systems need a bit of attention, too. If you install one in your system, you will need to periodically change oil and air filters. If you install a manually operated generator, you'll need to fire it up from time to time to raise the charge level on your batteries or to equalize the batteries. You may also have to haul your generator in for an occasional repair. Dan's backup generator has been to the repair shop twice in 13 years for costly repairs — and he only runs it 10 to 20 hours a year!

In grid-connected systems with battery backup, you'll have much less to worry about. If you install sealed batteries, for example,

Fig. 7.14: *Battery Filler Bottle. If you can access your batteries relatively easily, this filler bottle is one of the easiest and most economic means of adding distilled or deionized water to them.*

you'll never need to check the fluid levels or fill batteries.

Batteries may seem complicated and difficult to get along with, but if you understand the rules of the road, you can live peacefully with these gentle giants and get lots of years of service. Break the rules and, well, you're going to pay for your carelessness.

CHAPTER 8

INVERTERS

The inverter is a modern marvel of electronic wizardry that's an indispensable component of virtually all electric-generating renewable energy systems. Like batteries in off-grid systems, the inverter works day and night. Its main function is to convert DC electricity generated by wind turbines and solar electric modules into AC electricity, the type used in homes, offices, cabins and barns. But that's not all. Inverters in some systems perform a host of additional useful functions.

In this chapter, we'll take a look at the modern inverter. We'll examine the role inverters play in wind systems and then will take a peek inside this remarkable device to see how it operates. We'll also describe the features you should look for when purchasing an inverter. Before we delve into the subject, let's start with an important question: do you need an inverter?

Although this may seem like a ridiculous question, it's not. Some applications operate solely on DC power and don't require inverters. These include small wind or hybrid wind/ solar electric systems that power a few DC circuits, for example, in cabins or cottages. It also includes direct water-pumping systems that use AC or DC electricity from a renewable energy source to power an AC or DC water pump directly.

All other renewable energy systems require an inverter, but the type of inverter one needs depends on the type of system one has (see Chapter 4). Off-grid systems require an inverter designed to operate with batteries. Grid-connected wind electric systems require a utility-compatible inverter. Grid-connected systems with battery backup require a dual-purpose inverter — one that's grid- and battery-compatible.

Battery-Based Inverters for Off-Grid Systems

To understand how inverters operate, let's start with inverters used in off-grid systems, known as battery-based inverters.

Battery-based inverters perform two essential functions: (1) they convert DC to AC electricity and (2) they increase the voltage. In off-grid systems, electricity from the wind turbine is first fed into the batteries via a charge controller. When electricity is required, it flows from the batteries to the inverter. As noted in Chapter 7, the batteries are wired so they receive, store and supply 12-, 24- or 48-volt electricity. In a 24-volt system, then, the inverter converts 24-volt DC to 120-volt or 240-volt AC electricity — either 50 or 60 hertz, depending on the region of the world you're in. (In North America, AC is 60 hertz; in Europe and Asia, it's 50 hertz.)

As you shall soon see, the DC-to-AC conversion and the increase in voltage are performed by two separate but integrated components in the inverter. The conversion of DC to AC occurs in an electronic circuit known as a high-powered oscillator or a power bridge. It is also commonly referred to as an H-bridge. The second process, the increase in voltage, is performed by a transformer, a common electrical device. With this brief introduction, let's examine the conversion of DC to AC electricity.

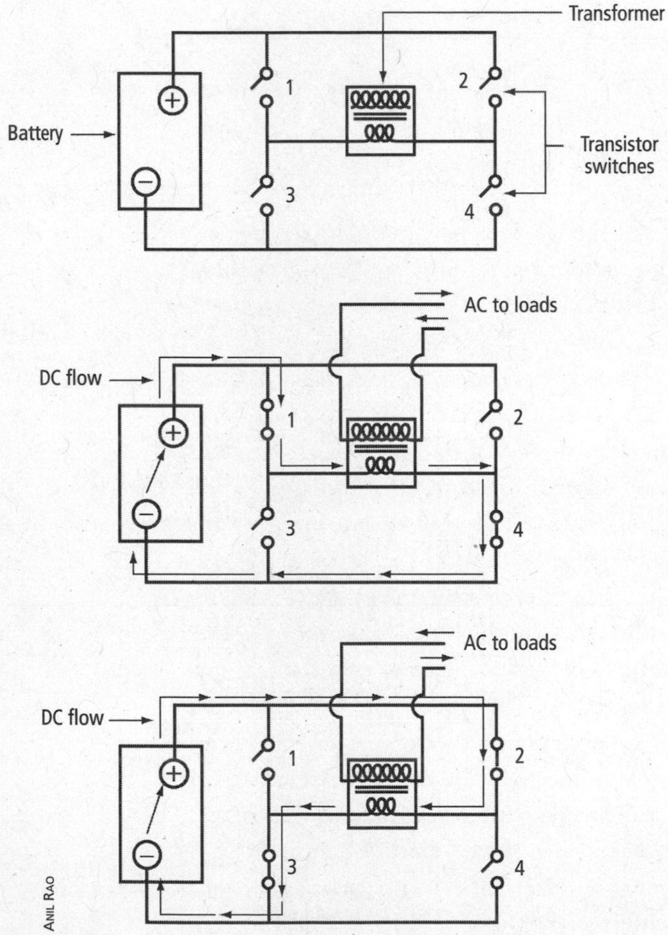

The Conversion of DC to AC

As shown in Figure 8.1a, the H-bridge consists of two vertical legs. Each leg contains two transistor switches. As shown in Figure 8.1b, DC electricity from the battery bank flows into the inverter. The arrows in Figure 8.1b show the initial direction electricity flows, which is controlled by the switches. When switches 1 and 4 are closed, DC electricity flows from the battery into the vertical leg on the left side of the H-bridge. The low-voltage DC electricity then flows through the horizontal portion of the H-bridge and out the bottom of the right leg of the H-bridge, and back to the battery.

Fig. 8.1: *Inner Workings of an Inverter. See text for an explanation.*

ANIL RAO

An electronic timer then flips the switches, as shown in Figure 8.1c. This allows electricity to flow from the battery into the vertical leg of the H-bridge on the right side of the diagram. Electricity then flows through the horizontal portion of the H-bridge in the opposite direction, out the vertical leg on the left side, and back to the battery.

Opening and closing the switches very rapidly allows DC electricity to flow first in one direction and then the other through the horizontal portion of the H-bridge and transformer. This produces alternating current at the same voltage as the battery bank. A small controller (not included in the figures) controls the opening and closing of the switches in the H-bridge.

Increasing the Voltage

Now that you see how an inverter produces alternating current in the central portion of the H-bridge, let's see how it boosts the voltage.

As illustrated in Figure 8.1, the horizontal portion of the H-bridge runs through one winding of a step-up transformer. A step-up transformer is a simple electrical device that increases the voltage of electricity. If you're unfamiliar with its operation, you may want to read the box on page 208.

The transformer in an inverter increases the 24-volt alternating current flowing through one winding to 120-volt alternating current in its other winding. The label "AC to loads" in Figure 8.1 b and c indicates the 120-volt AC electricity that flows out of the inverter. Although most inverters use transformers to increase the voltage of their AC output, others use electornic voltage boost circuits.

Square Wave, Modified Square Wave, and Sine Wave Electricity

An inverter with an H-bridge circuit produces alternating current electricity, but it's very choppy, as shown in the graph in Figure 8.3a. This graph plots the voltage over time. In alternating current electrons flow back and forth in a wire. If you could measure the voltage in a wire carrying AC electricity, you'd see that it changes over time. When the electrons are flowing one way, say to the right, the voltage climbs very rapidly to +120 volts. When the flow of electrons shifts direction, the voltage drops very rapidly to −120 volts. (The positive and negative signs refer to the direction of flow.) The graph in Figure 8.3a indicates the voltage over a very short period of time.

The basic H-bridge found in some older inverters produces an almost perfect square wave, which is not a very usable form of electricity. If this square wave is modified so that it pauses at zero volts briefly before reversing direction, the result is modified square wave electricity shown in Figure 8.3b. Although this is closer to a sine wave, the kind used in most homes, it is still not suitable for use in homes and businesses.

The problem with square wave — and to a lesser extent modified square wave — electricity is that they contain harmonics. Harmonics are waveforms at frequencies that are multiples of the desired frequency. Square wave electricity contains odd harmonics — that is,

The Scoop on Transformers

Transformers are used to increase (step up) and decrease (step down) voltage. If you're tied to the electrical grid, chances are there's a transformer on an electric pole outside your home (Figure 8.2). This device is known as a step-down transformer because it decreases the voltage of the electricity from 25,000 volts in the electric line running by your home to 120 or 240 volts, so that the electricity arriving in your home doesn't burn up your 120- and 240-volt appliances.

In battery-based inverters, transformers do just the opposite: they increase voltage. As a result, they are known as step-up transformers. Both step-up and step-down transformers use electromagnetic induction.

As explained in Chapter 5, electromagnetic induction is the production of an electric current in a wire as the wire moves through a magnetic field. However, magnetic fields are also produced as electricity flows through conductors. In conductors carrying alternating current, for instance, the magnetic field expands, collapses, and then reverses direction as the flow of electrons alternates. This alternating magnetic field can induce an alternating current in a nearby wire, or coil.

Transformers consist of two wire coils, called windings, one that's fed electricity, and another that produces electricity at a higher or lower voltage. AC electricity flowing through the first coil creates an oscillating magnetic field that induces a current in the second coil. How does a transformer increase or decrease voltage?

The voltage in the second coil increases and decreases in proportion to the number of turns in the second coil compared to the number in the first coil.[1] The greater the number of turns, the higher the voltage. If the number of turns in the second coil is greater than the first coil, voltage in the second coil increases. If the number of turns in the second coil is lower, the voltage in the second coil is lower.

Fig. 8.2: *Step Down Transformer. High voltage electricity travels along electrical lines. Step-down transformers reduce the voltage so that it enters our homes and businesses at 120 and 240 volts, which is required by most household appliances and electronic devices.*

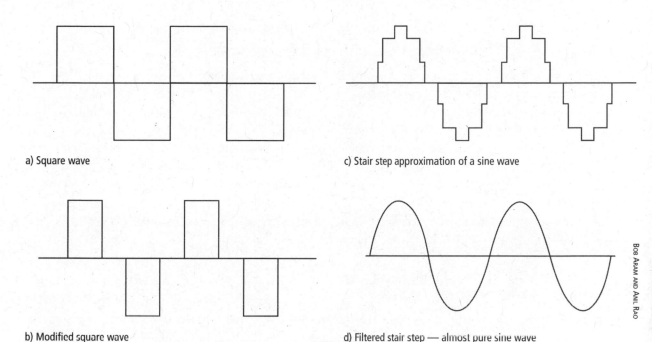

a) Square wave

c) Stair step approximation of a sine wave

b) Modified square wave

d) Filtered stair step — almost pure sine wave

BOB ARAM AND ANIL RAO

frequencies that are three times, five times, seven times, etc. the fundamental frequency of the square wave. Therefore, the output of a 60 hertz square wave or modified square wave inverter contains not only 60 hertz, but also 180 hertz, 300 hertz, 420 hertz, etc. power. Motors, transformers and many electronic devices don't perform well under such circumstances and can be damaged by this phenomenon, called harmonic distortion (see sidebar).

Filtering the output of a square wave or modified square wave inverter reduces these unwanted harmonics, but won't eliminate them.

Modified square wave electricity works OK in most applications, but it is not suitable for all electronic equipment. Moreover, most devices run less efficiently on modified square wave electricity. Because of this, most inverters sold in North America and Europe are designed to produce a much purer form of electricity known in the trade as sine wave electricity. This is nearly identical to the electricity delivered via the utility grid.

In sine wave inverters, multiple H-bridges, each of which produces square wave electricity, are stacked on top of one another — that is, carefully spaced to produce a stair step waveform that more closely resembles a sine wave, as shown in Figure 8.3c. Filtering this waveform produces a very clean sine wave, the kind of electricity modern homes require (Figure 8.3d).

Fig. 8.3: *Wave Forms (a) Square Wave, (b) Modified Square Wave, (c) Stair Step, and (d) Sine Wave Electricity. See text for discussion.*

What is Total Harmonic Distortion?

Total harmonic distortion (THD) is a measure of the harmonics in an inverter output — the frequencies we don't want — compared to the frequency we do want. Electricity from a square wave inverter has a THD of 48 percent. The output of a modified square wave inverter will be somewhat less than this, although you won't find a value on the inverter specification sheet. Inverter manufacturers don't like to talk about the THD of their modified square wave inverters and don't put this data in their spec sheets.

In sharp contrast, the THD of the output of a sine wave is five percent or less, which is about the same as the distortion in utility power. In fact, it can be better than utility power; depending on where you live and what your neighbors — and the utility — are doing that creates distortion on the power lines. A perfect sine wave has zero percent THD.

What Else Does a Battery-Based Inverter Do?

Battery-based inverters perform a host of other functions, besides those just described. Let's take a look at them. As you read this, bear in mind that this information pertains to inverters for off-grid systems and grid-connected systems with battery backup.

Battery Charging

Battery-based inverters typically contain battery chargers. (Batteryless grid-tied inverters don't.) Battery chargers are required to charge batteries from an external source — either the grid or a gen-set. But wait, you say, aren't the batteries charged by the wind turbine through the charge controller?

The charge controller in battery-based systems is not a batery charger, but a battery charge *regulator*. It regulates the charging of the battery bank from a wind turbine. It does not produce energy; it merely regulates the flow of the electricity from the wind generator to the battery bank. It also protects batteries from overcharging. A battery charger in the inverter, on the other hand, takes AC from the grid (in a grid-connected system) or a gen-set (in either an off-grid or grid-connected system) and converts it to DC. It then feeds DC electricity to the batteries.

As noted in Chapter 7, battery charging from gen-sets is used to restore battery charge after periods of deep discharge to prolong battery life and prevent irreparable damage. Battery chargers are also used to equalize batteries, also discussed in Chapter 7.

Abnormal Voltage Protection

High-quality battery-based inverters also contain programmable high- and low-voltage disconnects. These circuits protect various components of a system, such as the batteries, appliances and electronics in a home or business. They also protect the inverters. Let's take a look at the low-voltage disconnect first.

The low-voltage disconnect in an inverter monitors battery voltage at all times. When extremely low battery voltage is detected, indicating the batteries are deeply discharged, the inverter shuts off and/or sounds an alarm. The flow of electricity from the batteries to the inverter stops. The inverter stays off until the batteries are recharged.

Low-voltage disconnect protect batteries from very deep discharging — discharging to a level that could damage the battery (below the 80 percent mark). As noted in Chapter 7, although lead acid batteries are designed to withstand deep discharges, discharging batteries beyond 80 percent causes irreparable damage that leads to their early demise. Although complete system shutdown can be a nuisance, it is vital to the long-term health of a battery bank.

Although the low-voltage disconnect feature is critical, Ian notes, "It should not be used to manage one's batteries. It is designed more to protect inverters from low voltage than to protect batteries from deep discharge. If the voltage of the battery bank reaches the disconnect level, you may have already damaged your batteries." It's much better to develop an awareness of the state of charge of your batteries, and modify usage or turn on a generator well before the low voltage disconnect kicks in (as explained in Chapter 7).

To avoid the hassle of having to manually start a generator when batteries are deeply discharged, some inverters contain a circuit that activates the generator automatically when low battery voltage is detected. In these systems,

A Cautionary Note

Ian notes that some industry experts are leery about using auto-start generators, preferring instead oversight of a human who monitors the batteries and the condition of the generator, its fuel supply, and so on. One reason for this is that if the inverter malfunctions, for example, it starts the gen-set but doesn't stop it when the batteries are fully charged, the batteries can be overcharged and ruined. If your generator doesn't come with a low-oil protection, which shuts the machine off if the oil level is low, automatic start can ruin a generator.

the inverter sends a signal to start the generator, provided the gen-set has a remote start capability. The gen-set then recharges the batteries. As you might suspect, more sophisticated auto-start generators cost more than the standard pull-cord type. When the generator is operating, the inverter may also send some of the AC electricity it produces to the main service panel to service active loads or to a backup load panel (in grid-connected systems), also to supply active loads.

In grid-connected systems with battery backup, the grid serves the same function as the backup generator, although some grid-connected systems with battery backup include gen-sets to supply electricity in the event of a power outage.

High-Voltage Protection

Inverters in battery-based systems also often contain a high-voltage disconnect. This circuitry helps to protect the inverter from excessive battery voltage. It also protects the batteries from overcharging when the inverter's battery charger is operating. It stops the flow of electricity from the gen-set in off-grid systems or the utility grid in grid-connected systems when the battery voltage is extremely high. (Remember: high battery voltages indicate that the batteries are full.) High-voltage protection prevents overcharging, which can severely damage the lead plates in batteries.

Batteryless Grid-Tied Inverters

Now that you've seen how a battery-based inverter operates, let's take a look at batteryless grid-tied inverters, the most commonly used inverter in wind and solar electric systems today (Figure 8.4). These systems require inverters that are designed to operate in synch with the utility grid. Batteryless grid-tied inverters are similar to battery-based inverters, but with some notable differences. The most important difference is that they require grid connection to operate. To understand why, let's take a look at how a grid-connected wind system operates.

As noted in Chapter 5, most modern wind turbines produce "wild AC" — alternating current electricity whose voltage and frequency vary with the rotational speed (revolutions per minute) of the blades. The rpm of the blades, of course, varies with the wind speed.

Fig. 8.4: *Grid-Connected Inverter. This inverter by Magnetek is designed for batteryless grid-connected wind energy systems.*

In batteryless grid-connected systems, wild AC from the wind turbine is first converted to DC electricity by rectifiers either in a controller or the inverter. The inverter converts the DC output of the rectifier to AC electricity. To be certain that the AC output is in synch with electricity on the grid, batteryless inverters monitor the frequency and voltage of utility power. They adjust their AC output so that it matches the electricity being supplied by the utility. The synchronized 120-volt AC electricity is fed into household circuits. Surplus, if any, is back fed onto the utility grid. (Battery-based inverters in off-grid systems do not require synchronization.)

Another key difference between batteryless grid-tied and battery-based inverters is apparent when you buy a system. Although you'll have to shop around for an inverter for an off-grid system, when installing a wind turbine in a grid-connected system, the manufacturer does the shopping for you. In fact, buyers usually have no choice when it comes to grid-tied inverters for wind energy systems. Each manufacturer recommends inverters for their wind turbines, and typically sells the turbine and inverter as a package.

Manufacturers specify the grid-tied inverters for their wind machines because every wind machine has a different output voltage range. One turbine may, for instance, produce AC that ranges from zero to 300 volts. Another may produce wild AC from zero to 200 volts. Manufacturers select inverters whose input range corresponds to the output voltage of the turbine. If you purchase a Bergey Excel (21-foot rotor, rated at 10 kilowatt-hours), for example, it comes with an inverter manufactured by Xantrex. The Proven WT 2500 is sold with an SMA 2500 Windy Boy inverter. Abundant Renewable Energy's ARE 442 is sold with two SMA Windy Boy 6000 inverters. Southwest Wind Power's Skystream 3.6 comes with its own inverter, located inside the body of the turbine with the alternator.

As noted in Chapter 4, grid-tied inverters are programmed to shut down automatically if there's an increase or decrease in the voltage or frequency of the utility power that falls outside the inverter's acceptable limits. In other words, if either the voltage or the frequency varies from the settings programmed into the inverter, the inverter shuts down and stays down until the grid power returns within the acceptable range.

Inverters in grid-connected systems shut down if the grid goes down entirely or the frequency or voltage of grid power is out of spec (falls outside pre-set limits). The latter may occur at the end of a utility line or near a manufacturing plant that has a poor power factor.[2]

New Development in Grid-Connected Inverters

One of the new developments in grid-connected inverters is the emergence of a new standard (UL 141) that ensures safe interconnection of distributed generation sources. Most inverters today are UL 141 certified, which should smooth the interconnection process with any local utility.

In such instances, grid power is poor enough that it doesn't match the window required by the inverter.

When the inverter shuts down, the flow of electricity to the main service panel and to the utility both cease. In most systems the electrical output of the wind turbine is diverted to a dump load. In others the controller shuts down the turbine. Which of these occurs depends on the system design.

Why go through all this? As noted in Chapter 3, grid-tied inverters terminate the flow of electricity to the grid when there's a power outage to protect utility workers who may be working on the lines from electrical shock. Unfortunately, when the inverter shuts down the flow of electricity to the main service panel and active circuits is also terminated.

In grid-connected systems with battery backup, however, a homeowner or business can continue to operate. Although inverters in such systems disconnect from the utility if the utility goes down, the inverter continues to feed electricity to active circuits through a separate breaker box, a critical load panel. It is typically wired only to essential circuits in a home or business. Although a homeowner or business owner may see a momentary power flicker when the grid goes down, the better the inverter, the smoother the transfer. High-quality inverters switch so fast that even computers are not affected during the transfer.

Inverters designed for batteryless grid-connected systems may cost a little less than their more complicated cousins, off-grid inverters, depending on the model and manufacturer.

However, the real savings in a batteryless grid-connected system comes from the omission of batteries and a battery room. Additional savings result from the lack of battery replacement in subsequent years, as well as savings on the time required to maintain batteries, including periodic equalization. In new construction, however, when comparing costs of off-grid to grid-connected systems, don't forget to include line extension and utility connection costs incurred with utility-intertie systems.

Multifunction Inverters

The third type of inverter is a multifunction inverter, one that is used on grid-connected systems with battery backup. These inverters are popular among individuals with homes or offices who can't afford to be without electricity for a moment in their lives. They are also crucial for businesses. These applications require multifunction inverters — battery- and grid-compatible sine wave inverters. They're commonly referred to as multifunction inverters. Both applications require a charge controller.

Multifunction inverters combine features of both types of inverters — grid-connected and off-grid. On the grid-connected side, they contain islanding protection that automatically disconnects the inverter from the grid in case of loss of grid power, over/under voltage, and over/under frequency. They also contain fault condition reset — to power up the inverter when the problem with the utility grid is rectified. On the off-grid side, they contain battery chargers and high and low-voltage disconnects.

Grid-connected systems with battery backup are the most difficult to understand. Many people think that the batteries are used to supply electricity at night or during periods of excess demand — that is, when household loads exceed demand. That's not true. The grid is the source of electricity at night or during periods when demand exceeds the output of the wind turbine. The batteries are there primarily in case the grid goes down.

How does electricity flow during the day?

That depends on the state of charge of the batteries.

If the batteries are full, DC electricity from the wind turbine travels to the charge controller, then to the inverter to meet household demand. Surplus electricity is back fed onto the grid.

If the batteries need to be charged, for example, after a utility outage, the inverter powers loads from electricity drawn from the utility lines while simultaneously charging the batteries. If the wind is blowing, electricity from the turbine will also be used to charge the batteries. Once the battery bank is full, the system returns to normal operation.

At night, the home is supplied by AC electricity from the utility grid, not the batteries.

Multifunction inverters are not the most efficient inverters, but they allow system flexibility that utility-intertie inverters do not. That is, they allow the incorporation of batteries. If you are installing an off-grid system, you may want to consider installing a multifunction inverter in case you may want to connect to the grid in the future.

While grid-connected systems with battery backup may seem like a good idea, they do have some disadvantages. One of the most significant is that some portion of the electricity generated in such a system is used to keep the batteries topped off. This may only require a few percent but over time, a few percent adds up. In systems with poorly designed inverters, the electricity required to maintain the battery bank can be quite substantial.

Xantrex SW-series inverters, according to wind energy expert Mick Sagrillo are not optimized for grid sell back. They focus on keeping the batteries full. Back feed is a second priority. The homeowner pays for this in reduced system performance. For best results, we recommend inverters that prioritize the delivery of surplus to the grid while preventing deep discharge of the battery bank. "OutBack's inverters are a big step forward in this respect," notes Ian Woofenden, "and other manufacturers (including Xantrex's newer models — their XW series) are improving their products as well."

If you want the security of battery backup in a grid-connected system, we suggest that you isolate and power only critical loads from the battery bank. This minimizes the size of your battery bank, reduces system losses, and reduces costs. Unless you suffer frequent or sustained utility outages, a grid-connected system without batteries usually makes much more sense from both an economic and environmental perspective.

With these basics in mind, we turn our attention to the purchase of an inverter. Since inverters and turbines for batteryless grid-tied systems are typically sold as a package, most of what follows pertains to battery-based inverters.

Buying an Inverter

Inverters come in many shapes, sizes and prices. The smallest inverters, referred to as pocket inverters, range from 50 to 200 watts. They are ideal for small loads such as VCRs, computers, radios, televisions, and the like. Larger units range from 1,000 to 5,500 watts. Most homes and small businesses require inverters in the 2,500 to 5,500-watt range. Which inverter should you select?

Type of Inverter

The first consideration when shopping for an inverter is the type of system you are installing. As noted earlier, if you are installing an off-grid system, you'll need a battery-based inverter.[3] You will also need to purchase a wind turbine designed for battery charging — one that produces DC electricity at the same voltage as your battery bank. The wind turbine will also come with a controller.

If you are installing a batteryless grid-connected system, you'll need a turbine designed for this purpose and its matching inverter. As noted earlier, the inverter must be matched to the output of the wind generator. Inverters are usually sold with the turbine to ensure compatibility. If not, the manufacturer will sell you the wind machine and controller and will recommend a compatible inverter. Follow their advice.

If you are installing a grid-connected system with battery backup, purchase an inverter

that is both grid- and battery-compatible. Battery-based grid-tie inverters may also be a good choice for off-grid systems, in case you decide to connect to the electrical grid at a later date. Two such inverters are the OutBack GFX or the Xantrex XW series inverters (they replaced the SW series starting in 2007).

When purchasing an inverter through an installer, your choices may be limited. Most installers carry a line of inverters they are familiar with and have a high degree of confidence in. An installer will very likely also be a dealer for a line of inverters. If this is the case, the installer will make a recommendation that fits your needs from the inverters he or she carries, saving you the anguish of selecting one yourself. We should point out that there are not that many battery-based inverters available in North America and only certain equipment is certified under some states' incentive programs.

Even though a local supplier/installer may recommend an inverter, we urge you to read the following material. It will expand your understanding of inverters, and will help you better manage and troubleshoot your renewable energy system. If you are designing and installing your own system, which we recommend only after serious study and a couple of hands-on workshops, you'll surely want to know as much as you can about inverters. Here's what you need to know and consider as you shop for a battery-based inverter.

System Voltage

When shopping for a battery-based inverter, you'll need to select one whose input voltage corresponds with the battery voltage of your system. System voltage refers to the voltage of the electricity produced by a renewable energy technology — a wind turbine, a solar electric array, or micro hydro generator. That is, the generators in these machines are wired to produce either 12-, 24- or 48-volt electricity. The batteries are wired similarly. Inverters, in turn, increase the voltage to the 120 or 240 volts AC used in homes and businesses. Because all components of an off-grid renewable energy system must operate at the same voltage, the inverter must match the source (wind turbine) and batteries (if any). A 24-volt inverter won't work in a 48-volt system. If you are installing a 48-volt Bergey XL-R, you'll need a 48-volt battery-based inverter and must wire your battery bank for 48-volts. If installing a 24-volt Kestrel 1000, you'll need a 24-volt battery-based inverter and battery bank. It is always a good idea to talk with the wind turbine manufacturer's technical staff to obtain their input on the best inverter.

Modified Square Wave vs. Sine Wave

The next inverter selection criterion is the output waveform. As noted earlier in this chapter, battery-based inverters are available in modified square wave (often called modified sine wave) and sine wave. Grid-connected inverters are all sine wave so their output matches utility power.

Modified square wave electricity is a crude approximation of grid power that works fairly well in many appliances and electrical devices in our homes, including TVs, lights, stereos, computers, inkjet printers, power tools, refrigerators and washing machines. Although all these devices can operate on this lower quality waveform, they all run less efficiently, producing more heat and less of what you want — i.e., light, water pumped, etc. — for a given input.

So why do manufacturers produce modified square wave inverters?

The most important reason is cost. Modified square wave inverters are much cheaper than sine wave inverters. Expect to pay 30 to 50 percent less for one.

Another reason for the continued production of modified square wave is that they are hardy. They work hard for many years with very little, if any, maintenance. (Their durability

Wind Machine Options

Wind machines designed for battery charging typically come in one of three standard voltages: 12, 24 and 48 volts. The Proven WT 2.5, for instance, is available in 12-, 24- and 48-volt DC battery charging configurations, as is the Kestrel 1000, now manufactured by Eveready. Some models offer fewer options. The Kestrel 3000, for example, is available only in a 48-volt model. The Bergey XL.1 is only available in a 24-volt model. You may even encounter wind turbines designed for 36-volt systems, which are uncommon. Whisper 100 and Whisper 200 from Southwest Windpower come in 12-, 24-, 36- and 48-volt DC models. You may also notice that larger wind machines are only available in higher voltage. Bergey's Excel, for example, is available in a 48-volt model as well as 120- and 240-volt DC models for battery charging.

Third World Electricity

Ian likes to call modified square wave output "third world electricity." It's as good as or better than the utility grid electricity supplied in countries like the Dominican Republic and remote locations in Africa. Electricity customers there are used to fluctuating grid quality (and availability) and appliances used there may be more "sympathetic" to this poor quality electricity. But in modern countries with cleaner grids, appliances should be fed sine wave electricity.

may be related to their simplicity: they are much less complex than sine wave inverters.)

Modified square wave inverters come in two varieties: high frequency switching and low frequency switching units. High-frequency switching units are the cheapest of the two. They are also much lighter than low frequency switching models, and are, therefore, easier to install.

Low frequency switching modified square wave inverters cost more and weigh more than their high-frequency switching cousins, but are well worth the investment. One reason for this is that they typically have a much higher surge capacity. That means they can deliver greater surges of power needed to start certain electrical devices such as power tools, dishwashers, washing machines and refrigerators. More on surge capacity shortly.

Unless money is tight, we recommend sine wave battery-based inverters for off-grid systems. Their output is well suited for use in modern

homes with all their sensitive electronic equipment. Dan found that his energy-efficient front-loading Frigidaire Gallery washing machine would not run on modified square wave electricity. The microprocessor that controls this washing machine — and other similar models (except the Staber) — simply can't operate on this inferior form of electricity. Since he replaced this inverter with a sine wave inverter, he's had no troubles whatsoever.

Certain laser printers may also perform poorly with modified square wave electricity. The same goes for some battery tool chargers, ceiling fans and dimmer switches.

Making matters worse, some electronic equipment such as TVs and stereos give off an annoying high-pitched buzz or hum when operating on modified square wave electricity. Modified square wave electricity may also produce annoying lines on TV sets, and can damage sensitive electronic equipment.

When operated on modified square wave electricity, microwave ovens cook slower. Equipment and appliances also run warmer and might last fewer years on modified square wave electricity. Computers and other digital devices operate with more errors and crashes. Digital clocks don't maintain their settings as well. Motors don't always operate at their intended speeds.

With all of these disadvantages, it is no wonder that the modified square wave inverter has lost market share to the sine wave inverter. SMA, Xantrex and OutBack all produce excellent sine wave inverters at a reasonable price.

Output Power, Surge Capacity and Efficiency

When selecting an inverter, even a grid-tied inverter, you will need to consider three additional factors: continuous output, surge capacity and efficiency. Let's look at each one.

Continuous Output

Continuous output is a measure of the power an inverter can produce on a continuous basis — provided there's enough energy available in the system. The power output of an inverter is measured in watts, although some inverter spec sheets also list continuous output in amps (to convert simply use the formula watts = amps x volts). Xantrex's sine wave inverter, model SW2524, for instance, produces 2,500 watts of continuous power, which means the unit can power a microwave using 1,000 watts, an electric hair dryer using 1,200 watts, and several smaller loads simultaneously without a hitch. (By the way, the 25 in the model number indicates the unit's continuous power output; it stands for 2,500 watts. The 24 indicates that this model is designed for a 24-volt renewable energy system.) The spec sheet on this inverter lists the continuous output as 21 amps.

OutBack's sine wave inverter VFX3524 produces 3,500 watts of continuous power and is designed for use in 24-volt systems. Most homes can easily get by on a 2,500- to 3,500-watt or, at most, a 4000-watt inverter.

To determine the continuous output you'll need, add up the wattages of the common appliances you think will be operating at once. Be reasonable, though. Typically, only two or three large loads operate simultaneously. When calculating continuous power, add 20 to 25 percent to the wattage listed on large induction motors like those found in central air conditioners and many woodworking tools.

Surge Capacity

Electrical devices with motors, such as vacuum cleaners, refrigerators, washing machines and power tools, require a surge of power to start up that typically lasts only a fraction of a second. Even though the power surge is brief, if a system can't provide sufficient power, the power tool or appliance won't start. And not starting is bad — the stalled motor will draw excessive current and overheat very fast. Unless it is protected with a thermal cut out, it may burn out.

When shopping for an inverter, be sure to check out the surge capacity. All quality low-frequency inverters are designed to permit a large surge of power over a short period, usually around five seconds, which greatly exceeds their continuous output. Surge capacity or surge power is listed on spec sheets in either watts or amps.

Efficiency

Converting one form of energy to another results in a loss of energy. Efficiency is calculated by dividing the energy coming out by the energy going in. When you burn a gallon of gasoline in your car, for instance, you only get about 10 to 20 percent of the chemical energy contained in the gasoline back out in the form of useful work — motion. So the efficiency of a car is 10 to 20 percent. The rest of the energy in the gasoline is lost in the conversion of chemical energy to heat. Energy losses resulting from such conversions are a fact of life.

Electrical conversions occurring in inverters are no exception to this rule. Fortunately, efficiency losses in inverters are quite low — usually only 5 to at the most 15 percent. It should be noted, however, that inverter efficiency varies with load. Generally, an inverter doesn't achieve its highest efficiency until output reaches 20 to 30 percent of its rated capacity, according to Richard Perez, author, renewable energy expert, and publisher of *Home Power* magazine. A 4,000-watt inverter, for instance, will be most efficient at outputs between 800 and 1,200 watts. At lower outputs, efficiency is dramatically reduced.

Cooling an Inverter

Inverters produce heat internally. This heat comes from the energy lost in the conversion process. A 4,000-watt inverter running at rated output at 85 percent efficiency produces 600 watts of heat internally. The inverter must get rid of this heat to prevent damage.

Inverters rely on cooling fins and fans to dissipate excess heat to reduce internal temperature. If the inverter is in a hot environment, however, it may be difficult for it to dissipate heat quickly enough to maintain a safe temperature.

To prevent damage, inverters are equipped with a circuit that reduces the electrical output as temperature rises. A Xantrex SW Series inverter, for example, produces 100 percent of its continuous power at 77°F, but only 60 percent

at 117.5°F. As the internal temperature of the power oscillator increases, the output current decreases. This means that if you need a lot of power and the internal temperature of your inverter is high, you won't get it.

The implications of this are many. First, inverters should be installed in relatively cool locations. Dan's is mounted in the coolest part of his earth-sheltered house, his utility room. If the inverter is mounted outside, don't put it where the sun will shine on it. Second, inverters should be installed so that air can move freely around them. Third, be sure to purchase an inverter that does a good job of dissipating heat. Ideally, you want an inverter that cools itself passively — one that doesn't require a cooling fan that consumes electricity. Many battery-based inverters have fans to cool them during battery charging via a gen-set.

OutBack's inverters have a die-cast aluminum casing with finned sides that dissipate internally generated heat passively (Figure 8.5). In some models, the internal circuitry is completely sealed off from the outside, protecting it from dust, moisture, insects and small rodents that "spell death to an ordinary inverter," according to the manufacturer. Their vented models, however, allow outside air to flow through the internal electronics, improving performance. These units produce more AC power than the sealed inverters and thus perform better in hot environments.

Battery Charger

Battery chargers are standard in most battery-based inverters installed in off-grid or grid-connected systems with battery backup. This allows you to charge your bank from the utility (for grid-connected systems) or a generator (for off-grid systems).

Noise and Other Considerations

Battery-based inverters are typically installed inside, close to the batteries to reduce line loss. Grid-tied inverters are almost always installed near the main circuit breaker panel where the utility service comes into the house. (Most inverter manufacturers like their equipment to be housed at room temperatures.)

If you are planning to install an inverter inside your home or office, be sure to check out the noise it produces. Inquire about this upfront; better yet, ask to listen to the model you are considering in operation. Don't take a manufacturer's word for it. Dan's Trace inverter (now manufactured by Xantrex) is described by the manufacturer as "quiet," but it emits a loud and annoying buzz. The first six months after he moved into his home, the inverter's buzz drove him nuts. Now he's used to it.

Some folks are also concerned about the potential health effects of extremely low frequency electromagnetic waves emitted by inverters as well as other electronic equipment and electrical wires. If you are concerned about this, install your inverter in a place away from people. Avoid locations in which people will be spending a lot of time — for example, don't install the inverter on the other side of a wall next to your bedroom or office.

While we're developing a checklist of features to consider when purchasing an inverter,

OUTBACK POWER SYSTEMS

Fig. 8.5: *The OutBack VFX series is a ventilated version of the company's original sealed FX series sine wave inverter/charger. The fluted sides dissipate heat generated internally. The inverter also allows cooler outside air to flow through the internal electronics, making more AC power available in extremely hot environments compared to the sealed FX series.*

be sure to add ease of programming. Dan's first Trace inverter (DR2424, modified square wave) was simple to program. All the controls were manually operated dials. To change a setting, all he had to do was turn the dial. His new Trace PS2524 is a sine wave inverter that works wonderfully (although noisily) in all respects. However, the digital programming is extremely complicated and the instructions are extremely difficult to follow.

Our advice on this subject is to find out in advance how easy it is to change settings, and don't rely on the opinions of salespeople or renewable energy geeks who can recite pi to the 20th decimal place. Ask friends or dealers/ installers for their opinions but also ask them to show you. You may even want to spend some time with the manual to see if it makes sense before you buy an inverter.

Another feature to look for in a battery-based inverter is power consumption under search mode. The search mode is an operation that allows battery-based inverters in renewable energy systems to shut down almost entirely when there are no active loads. The search mode saves energy because inverters can consume 10–30 watts or more from the battery just to remain on.

Although the inverter is sleeping when in search mode, it is not dead to the world. It's sleeping with one eye slightly open, on the alert should someone switch on a light or an appliance. How does it do this?

While in search mode, the inverter sends out tiny pulses of electricity approximately every second. They are sent through a house's electrical wires. When an appliance or light is turned on, the inverter senses the load and quickly snaps into action, powering up and feeding electricity to the device.

The search mode is handy in houses in which phantom loads have been eliminated. (As noted in Chapter 4, a phantom load is a device that continues to draw a small electrical current when off.) Eliminating phantom loads saves a small amount of electrical energy in an ordinary home. (According to the Department of Energy, phantom loads, on average, account for about five percent of a home's annual electrical consumption.) In a home powered by renewable energy, it saves even more because supplying phantom loads 24 hours a day requires an active inverter. An inverter may consume about 10 to 30 watts when operating at low capacity. Servicing a 12-watt amp phantom load, therefore, requires an additional 10 to 30 amp investment in energy in the inverter.

Saving energy by eliminating phantom loads and continuous inverter operation does have its downsides. For example, automatic garage door openers may have to be turned off for the inverter to go into search mode. When you arrive home late from a night out on the town, you'll have to get out of the car and switch on the breaker that controls the circuit that feeds the automatic garage door opener.

Another problem occurs when electronic devices like cell phone chargers that require tiny amounts of electricity are plugged in. When left plugged in, many cell phone chargers, for instance, draw enough power to cause an inverter to turn on. Once the inverter starts, however, the device doesn't draw enough power to keep the inverter going. As a result, the inverter switches on and off, *ad infinitum*. Dan found the same thing happened with his portable stereo.

Most modern homes have at least one load that's always on, for example, hard-wired smoke detectors, which require the continuous operation of the inverter. Because of this, many people turn the search mode off so that the inverter runs 24 hours a day.

Stackability

Finally, when buying a battery-based inverter, you may also want to select one that can be stacked — that is, connected to a second or third inverter of the same kind. Stacking permits homeowners to produce more electricity in case demands increase over time. Two inverters can be wired in parallel, for example, to double the amp output of a battery-based wind system.

A couple of inverter manufacturers such as OutBack and Xantrex produce power panels, easily mounted assemblies that house two or more inverters for stacking. Figure 8.6 shows a power panel by OutBack that contains two inverters. Xantrex's new XW inverters come with similar components.

Many homes require 240-volt electricity to operate appliances such as electric clothes dryers, electric stoves, central air conditioning or electric resistance heat. As a general rule, we recommend that you avoid such appliances, especially when installing an off-grid system. That's not because a wind or hybrid wind and solar system can't meet their needs, but rather because these appliances tend to use lots of electricity and you'll need a larger and more costly system to power them. Well-designed, energy-efficient homes can usually avoid using 240VAC, which also simplifies home and system wiring. A frequent exception is a deep well pump, which may require 240 volts, but in most cases, high efficiency 120VAC pumps, or even DC pumps, can do the job.

If you must have 240-volt AC electricity, you can purchase inverters that are designed to be stacked, in this case wired in series, to produce 240VAC, such as the OutBack FX-series inverters. Or you can purchase an inverter that produces 120- and 240-volt electricity, such as Magnum Energy's MS-AE 120/240V Series Inverter/Charger or one of Xantrex's new XW series inverters. Or you can install a step-up transformer such as the Xantrex/Trace T-240. This unit converts 120-volt AC electricity from an inverter to 240-volt AC. Or you can simply install a dedicated 240-volt output inverter for that load.

Conclusion

If you are buying an inverter, be sure that the supplier — be it an Internet supplier or a local supplier — takes the time to determine which inverter is the correct choice for your system. Ask lots of questions.

A good inverter is key to the success of a wind energy system, so shop carefully. Size it appropriately. Be sure to consider future

Fig. 8.6: *OutBack Power Panel.*

electrical needs. But don't forget that you can trim electrical consumption by installing efficient electronic devices and appliances. Efficiency is always cheaper than adding more capacity! When shopping, select the features you want and buy the best inverter you can afford. Although modified square wave inverters work for most applications, it is best to purchase a sine wave inverter.

CHAPTER 9

MAINTAINING A WIND-ELECTRIC SYSTEM

"Wind turbines are machines. They have moving parts," notes John Hippensteel owner of Lake Michigan Wind and Sun. "That means they require maintenance."

Maintenance takes time and money.

Just as your car, home and bicycle require occasional maintenance, so does a wind energy system. You wouldn't expect to drive your car for a year without changing the oil, rotating the tires, or performing other basic maintenance and repair. Ignoring these tasks would surely reduce performance and would very likely lead to even more costly maintenance in the future — or even catastrophic failure.

While regular maintenance is vital to cars, it is even more important for wind generators because they work longer and harder in more severe environments.

This chapter discusses maintenance of wind energy systems except for batteries, which are covered in Chapter 7. It is intended to apprise readers of a vital, though sometimes overlooked, part of living with a wind-electric system. This discussion is not meant to discourage you from pursuing wind energy, but rather to inform you of the maintenance requirements of wind energy upfront so you know what you're getting into. We want you to embark on this venture with your eyes wide open. Bear in mind that you can always hire a professional to inspect and maintain your system.

Buyer Beware

Don't be lured by exaggerated claims of maintenance-free operation made by wind turbine manufacturers or installers. No wind generator is "maintenance free," unless you leave it in its box.

What's Required

Most wind installers recommend an annual or biannual inspection. Although some wind turbines may be able to operate five years or more without maintenance or inspection, don't wait that long.

Mick recommends twice a year inspections of wind systems. The first is in the spring after the weather warms, but before thunderstorm season. Thunderstorms are likely the most violent weather your turbine will encounter, and you want to be sure it's in good shape before Nature's onslaught. The second inspection should be in the fall to be sure your turbine is ready for winter. The last thing you will want to do is climb a tower in a 30 mile-per-hour wind at 30 degrees below zero to take care of something that you could have fixed on a nice, warm fall day.

Inspections should be performed on windless days, of course. This makes climbing the tower (in the case of a freestanding or fixed guyed tower) or lowering the tower (in the case of a tilt-up tower) easier, safer and more comfortable. Wind turbines on climbable towers should be shut down to prevent injury. Never perform inspection and maintenance when the blades are spinning. We want you to be around to enjoy wind energy for a long time. As an added safety measure, after you've climbed the tower, tie the blades to the tower so they can't spin, just in case the winds start to blow while you're working on the turbine. Also, be sure to brake a wind turbine on a tilt-up tower before lowering the tower.

Inspections and maintenance vary with the wind machines and the towers on which they are mounted. Installation manuals that come with turbines and towers usually provide detailed instructions. Be sure to read them carefully, and call the manufacturer for any maintenance updates and advice. You may also want to attend the Midwest Renewable Energy Association's maintenance and repair workshop. Recognizing that each machine is different, we offer some general guidelines.

Climbable Towers

Before you climb the tower, listen for unusual sounds — for example, clunking, banging or grinding. They indicate a bad alternator or bearings. Check for vibrations. Actually, sounds coming from the wind turbine should be noted any time it is operating. The turbine should operate smoothly and quietly.

Also, before climbing a tower to check the turbine on a fixed guyed tower, inspect the guy cables for proper tension. Tighten, if necessary. Check the connections of guy cables to the anchors. Tighten loose nuts. Check the foundation and the anchor bolts. Be sure that the nuts are tight and are not rusted, or corroded, or cracked. If they are, replace them immediately.

When climbing a tower be sure to wear a safety harness (preferably a full-body harness). Secure all tools and extra parts so they don't fall and injure people or pets below you. Although we discussed safe climbing in Chapter 7, it's worth repeating some of that advice. First, all climbable towers should be fitted with an anti-fall cable, a fixed climbing cable. An anti-fall cam-type mechanism is attached to the safety harness to catch you should you slip while you are climbing. This sliding "climbing car" follows you as you ascend but locks onto the cable to arrest a fall. Once you are atop the tower, belt in with your lanyards and disconnect from the anti-fall cable. Be sure to always secure yourself to the tower. When climbing a tower without a safety cable, "Always climb using two lanyards in an alternating pattern so that one of them is clipped onto the tower at all times," advises Jim Green.

Be sure to read through the entire inspection and maintenance protocol *before* ascending the tower to be certain you have all the tools and parts you need. Make a checklist of tools and inspection and maintenance duties on the front and back of a note card. Carry it with you to the top of the tower. You might want to put a duplicate copy in your pocket in case a breeze comes and sends your list on a cross-country flight.

On your way up, inspect all tower hardware at each tower section joint and at each guy station, where guy cables attach to the tower. Carry a wire brush with you and a spray can of galvanizing compound. Use the brush to remove rust and then cover bare or rusted spots with the galvanizing compound. Replace any hardware that is questionable. If the cross-bracing for the tower is welded, check for cracks or rust at the welds as you ascend. Cracks or rust must be attended to immediately. Cracked welds indicate that something is seriously wrong — do not ignore them! Lower the tower and fix them ASAP.

If your tower is equipped with removable step-bolts for the stub tower (the last section of tower that connects to the turbine), be sure to carry them with you when you climb, using a closeable bag. Or better yet, ask your ground crew to raise them to you on a service line. This line can be a quarter-inch nylon cord run through a pulley at the tower top. The first climber carries the rope and pulley up and then securely attaches the pulley to the top of the tower. The rope and pulley can be used to raise and lower parts. If your tower is equipped with a work platform, check to be sure it is firmly attached to the tower.

When climbing a tower or working on a platform, be sure you are securely tied in or clipped in at all times. We recommend at least two lanyards to secure you to the tower once you've reached the top. As John Hippensteel says, one lanyard is for his wife and the other is for his children. Always employ a ground crew of at least one person who is there at all times to assist you and other members of the climbing crew, if any. This can save hours of climbing up and down the tower, and is imperative for safety.

When you've reached the top of the tower, check the condition of all bolts that attach the

Read Before Your Climb

Be sure to read through the *entire* inspection and maintenance protocol *before* ascending the tower to be certain you have all the tools and parts you need. Make a checklist of duties in order. Write big enough so that you can read the list easily. Print the list on one card that fits neatly into your pocket. Carry a spare in case the first one is dropped or swept away by a breeze.

turbine to the stub tower. Tighten if necessary. Replace if damaged. Then check the condition of the bolts that hold the blades in place. Tighten loose bolts. Next check the tail vane bolts (the bolts that attach the tale vane to the boom) and any other fasteners. Vibration is a way of life with wind generators, and it causes bolts to loosen over time. You may need to use a torque wrench to tighten bolts properly. Manufacturers include specifications in their installation manuals for proper tightening of all bolts. Be sure to include these torque specifications on your list.

Vibration can also crack metal at bolt holes. Pay particular attention to thin metal parts such as the tail vane. If cracks are present, make a note to order a replacement. Cracked parts on a wind turbine never heal themselves, regardless of how much time you give them.

The blades undergo the greatest wear and tear of any part on a wind turbine, regardless of whether they are made of fiberglass, plastic, wood or a composite. If your wind machine has painted blades, you may need to remove, re-sand and repaint or refinish the blades to protect them from the elements. (This is obviously a job to be performed on the ground.) If the paint has cracked or the leading edge tape has torn away, the exposed blade material will begin to erode due to abrasion. Moreover, moisture that penetrates the blade can cause the rotor to become unbalanced, severely stressing the wind generator. If your blades show considerable wear and tear, you may want to perform this maintenance more regularly, for example, every six months.

If your turbine has plastic blades, check for cracks, especially near the base of the blade — the point of connection to the hub. This is a very high-stress area. Check the condition of the leading edge tape, if any. If necessary, re-apply leading edge tape to protect blades from abrasion caused by blowing dust or insects. Check the tips and trailing edge for damage, too. If damaged, repair the blades immediately. After ten years, the blades may need to be replaced, depending on how attentive you were to the first signs of erosion on the blades. Leading edge tape and fiberglass used to repair a slightly damaged blade will look cheap compared to replacing a $6,000 set of blades!

While you are at it, check the blades for tracking by measuring the distance from the stub tower to the tip of each blade. You don't want the blades to strike the tower as they bend under wind loads. Contact the manufacturer for tolerance and shim blades at the root if necessary. (Shimming helps maintain the proper distance between the blades and the tower.)

After checking the blades, examine the alternator of the turbine. Look for cracks or damage on all parts of the generator or alternator. Damage here is rare, but it is worth looking for. Cracks in any part of an alternator are an indication that something is seriously wrong. Shut the wind generator down immediately. Replace the damaged part before running the wind turbine again.

Next, check the alternator brushes, if any. They should allow the rotor to move easily. (Most modern turbine alternators are brushless.) Check the alternator bearings to be sure the seal is intact and no grease has leaked out. Replacing a damaged bearing that is starting to rumble is far cheaper than replacing the entire alternator because the magnets and windings ground each other into oblivion.

Once you have inspected the blades and alternator for damage, it is time to inspect other internal components. Your examination will require a high degree of familiarity with its components, such as the slip rings. If you are unfamiliar with them, be sure to carefully study diagrams *before* you ascend the tower for inspection. (Here's where an experienced inspector who's familiar with your turbine make and model is worth his or her weight in gold!)

Begin by checking for obvious signs of corrosion. Check the slip rings for damage or for oil that may have leaked from yaw bearings, the component that allows the wind machine to turn on the tower as the wind changes direction. Inspect all parts for cracks or wear. In some wind machines, you may need to grease certain parts or change the oil in the gearbox, if your turbine has one. The Jacobs 31-20, for instance, requires a complete oil change each year.

After you have inspected the internal parts and closed the access panel or cover plates, check the furling and shut-down mechanisms. If a cable is used to manually fold the tail, inspect the cable. Check its attachment to the tail boom. Look for fraying (unraveling) of the cable. Next, check for cracks and loose nuts in the tail assembly and hinges. Be sure to check the tail pivot pin, pin retainer bolts, and tail pivot bushings (these are all parts that allow "furlable" tails to fold). Examine the surfaces that contact each other at each end of the furling range. These are common wear points, and may need paint, at a minimum.

If overspeed protection is provided by a vertical furling mechanism or a blade pitch mechanism, be sure to inspect the bolts, pins and springs carefully for damage. Tighten loose bolts as per manufacturers' recommendations and grease as needed.

With your inspections and maintenance complete, it's time to head down the tower. As you descend, make sure that the electrical wire running from the wind machine to the junction box at the ground is secured to the tower. Replace fasteners and locking nuts as needed, especially any that have rusted or cracked. If your wind machine has a furling cable to shut the machine down for inspections or to protect it from high winds, check the condition of the cable as you descend.

When you have reached the base, be sure the furling winch, if any, is in good shape and operating correctly. Also, check the connections on ground rods. Be sure that all contact surfaces are clean and free of oxidation. Inspect other electrical components such as surge arrestors for damage and, if necessary, replace them. If your system is equipped with a dynamic brake, a mechanism that shorts the alternator to stop the turbine, be sure to check the switch.

Tilt-up Towers

If your turbine is mounted on a tilt-up tower, you will need to lower it to inspect the turbine. However, before you begin that procedure, check all of the guy cable attachments to the anchors. Check for loose or rusted cable clips or hardware and tighten or replace as needed. Check the tilting hinge at the tower base and the attachment of the gin pole to the tower. Make sure that the ground rods are securely attached to the guy cables.

After you lower the tilt-up tower, walk along the length of the tower to check the guy cable attachments to the tower. Tighten or replace cable clips or hardware as necessary.

Finally, check the wires coming out of the bottom of the tower to make sure that they are not chafing or chewed up by rodents. Once you are satisfied that the tower is in good shape, inspect the wind turbine as outlined for climbable towers.

Final Inspections

When the inspection is complete and you are off the tower (or your tower is tilted back up), you can turn the wind turbine back on. It's now time to check the inverter and/or controller to be sure they are working properly. Check for dust or moisture and eliminate the sources.

You also need to test the electrical production of the turbine. Detailed instructions should be available in your installation manual to ensure that the turbine is working properly. Contact the manufacturer if you have any questions on test procedures. Be sure to talk to engineers, not sales personnel.

Conclusion

Routine inspection and maintenance of a wind machine and tower is a kind of insurance policy.

Like taking your car in for oil changes and other scheduled maintenance, it helps ensure longer lifespan and can save you a lot of money over the long haul. Tightening a loose nut is a lot cheaper than replacing a part, or worse, the entire wind turbine that's been damaged when a nut comes off. Procrastinating could create a time bomb, a wind turbine that will decide its maintenance or replacement schedule for you.

Proper maintenance can help the system last for 20 to 30 years — helping you get the most for your investment. It will also help reduce the amount of mechanical sound a wind turbine produces.

The tower and wind turbine are not all that require maintenance. If you install a battery bank, you'll need to maintain the batteries. Off-grid battery banks typically require a lot more care than batteries installed in grid-connected systems because the latter typically contain no-maintenance sealed batteries. Details of battery maintenance are discussed in Chapter 7.

In contrast, inverters are generally fairly maintenance free. The only attention they require is when they break down, which is rare.

As should be clear by now, routine maintenance and repair are essential to all wind energy systems. If you're not up to the task or can never seem to get around to chores, you may want to hire a professional to inspect your turbine, tower and other components every year. If that's not an option, you may want to think about a less maintenance-intensive renewable energy system, like a solar electric array.

CHAPTER 10

FINAL CONSIDERATIONS:

ZONING, PERMITS, CONVENANTS, UTILITY COMPANIES, INSURANCE AND BUYING A SYSTEM

To those of use who love the idea of generating electricity from the wind, there are few things, if any, in the world more rewarding than raising a tower with a wind turbine and watching the turbine's blades spin for the first time, generating electricity from a clean, abundant and, in some locations, an extremely reliable source of energy. The road from conception to installation of a turbine on a tall tower on your property, however, can be long and arduous. The steps are summarized in Table 10.1. As you can see, we have already covered the first three.

In this chapter, we'll discuss most of the remaining steps. We'll begin by discussing building permits and zoning issues, and then turn our attention to interconnection agreements with local utilities, a subject relevant to those who want to install a grid-connected system. Next, we'll discuss concerns neighbors might have and ways to work with them. We will also cover

insurance requirements. We'll conclude by providing advice on buying a system and locating a professional installer.

Zoning, Building Permits, and Restrictive Covenants

The decision to install a wind system often involves more than an economic analysis to determine if a wind generator can produce electricity cheaply enough to make the investment worthwhile. There are several additional matters you'll need to check into first. For one, you need to check into zoning regulations — potential legal restrictions imposed by local or state government. These may include height restrictions or setbacks — how far a turbine must be placed from property lines and utility lines.

Zoning Regulations

Zoning laws regulate how parcels of land can be used and are common in many parts of the

Table 10.1
Steps to Implement a Small Wind Energy Project

1. Measure your electrical consumption.
2. Assess your wind resource.
3. Select a turbine and tower.
4. Check zoning regulations and homeowner association regulations.
5. Check building permit and zoning requirements.
6. Check covenants.
7. Contact the local utility and sign utility interconnection agreement (for grid-connected systems).
8. Obtain building and electrical permits.
9. Order turbine, tower and balance of system.
10. Install system.
11. Commission — require installer to verify performance of the system.
12. Inspect and maintain the system on an annual basis.

world. Although zoning regulations are in place in numerous cities and towns in North America, some regions, specifically rural areas, have no zoning regulations whatsoever. You're free to do whatever you want. (These regions have the authority to zone, but don't do it.)

Wind turbine installations are typically subject to zoning laws, so *before* you purchase a system, be sure to check local zoning regulations to be certain it's legal to install a wind turbine, and, if so, what restrictions may be in place. For example, zoning regulations may stipulate tower height and setbacks.

In cities and towns with zoning, wind turbines may or may not be allowed. If they are allowed, you'll all too frequently find that zoning regulations limit tower height to 35 feet, which is useless when it comes to a wind generator.[1]

To obtain permission to install a taller tower, you'll need to obtain a variance or a special use permit — permission to legally vary from zoning regulations. These are issued one property at a time. (Developers receive variances all the time, so don't be daunted by this requirement.)

Obtaining a variance or special use permit may take several months and can cost thousands of dollars. You'll need to submit a formal request and attend a hearing and may need to hire an attorney to assist in the process. If others have been granted a variance to install towers in your jurisdiction, all the better. Legal precedent could make your job a lot easier. "Knowing someone on the city or town council, or even in the mayor's office, can take months, even years, off this process," notes Aaron Godwin, an installer of small wind energy systems.

One problem that applicants for residential wind turbines often come up against is that local zoning regulations do not have stipulations covering wind-electric systems. As a result, local officials may consider the request under other sections of local zoning regulations. "This is particularly the case when it comes to tower height," notes Mick in one of his many articles on zoning and building permits in *Windletter*. "Wind generator towers are sometimes treated like any other structure on a homeowner's property, and, unfortunately, often severely restricted in height." In many areas, zoning regulations limit building height and, by default tower height, to 35 feet.

Building height restrictions have been on the books forever! And many code officials have no idea why. We're told they emerged many years ago to protect factory workers jammed into substandard factories — wooden structures with poor electrical wiring. Because of limitations of early fire-fighting equipment, which could only pump water about 32 feet, early codes set a 35-foot height for buildings. This number was grandfathered into many building codes and still exists around the United States, according to Mick.

Mick further notes that variances are often granted to permit construction of tall silos, radio and transmission towers, and cell phone towers. "Wind turbine towers," Mick asserts, "should also be exempted from height restrictions as they become ineffectual at or below the tree line." He adds, "Once zoning committees understand the reasons behind tall towers for residential wind systems, most will entertain height exemptions."

Building and Electrical Permits

Most people who want to install a wind turbine in the United States and Canada are required to obtain a building permit. Building permits ensure that the project (1) is safe, (2) is on your property and within required setbacks, and (3) complies with local ordinances, including zoning, if any. They also ensure that your improvement is added to your property records for property tax assessment. Building permits come with a cost. Fees vary from $50 to $6,000, depending on the jurisdiction. How do you find out about potential legal hurdles like permits?

"The quickest way to determine the local codes and requirements is to call or visit the office of the building inspector," note the folks at Bergey Windpower.

Building permits are issued after a review of plans that include a site map, drawn to scale, that indicates property lines and dwellings, including nearest neighbors and the location of other buildings. They also show topographic features, easements, if any, and the location and height of the proposed wind machine.

Permit applications also indicate the kind of turbine and tower you'll be installing as well as the tower's height and location. You may also have to submit data on sound production. If the zoning regulations require setbacks — placement of a tower a certain distance from property lines, streets, and overhead utility lines — the site plan must indicate your compliance with them. Setbacks are required in many cases so that if a tower collapses, it will fall within the bounds of your property. (Tower collapse is extremely rare and greatly blown out of proportion.)

Although setbacks can affect the placement of a wind turbine, don't let them deter you. "Few 600-foot television transmitter towers are smack in the middle of a 1,200- square-foot lot and utility transmission towers are only required to have a narrow right-of-way. If you can find any such exceptions in your area, you have a precedent in your favor," advises Mick. Better yet, he says, talk with your neighbors about setback. "Most ordinances are written so that if your neighbor has no objection to siting the tower closer to the property line, the permit can be granted."

Building permit applications often require drawings of the tower foundation and technical information on the wind turbine as well. Homeowners may rarely be required to submit an engineering analysis of the foundation and tower to demonstrate that both are structurally sound and comply with local building codes. Engineering analyses are typically provided by the turbine and tower manufacturer and may be stamped by their own engineer. (Their stamp is known as a "dry stamp.")

Some municipalities require a wet stamp — a stamp of approval by a state-licensed engineer, which could cost $500, possibly more. Check requirements in advance. If this is the case, you may want to appeal the requirement. As Mick points out, "Requirements for a 'wet stamp' cannot be justified. Any company selling towers for wind systems has to perform the engineering analysis to be in business and secure liability insurance." To learn more about tower engineering for building permits, see Mick's article in AWEA's *Windletter* (May 2006).

Even if not required, it is always useful to submit drawings of the tower and the tower footings when requesting a building permit for a wind system, notes Mick: "A good detailed blueprint will go a long way toward assuring zoning officials that your project is well thought out." Even though the blueprint may not state it, you can assure the building department officials that commercially available towers conform to the requirements of the Uniform Building Code, often referred to as the UBC. If not, they would not be sold by the company.

Electrical permits may also be required to ensure the system is installed according to local electrical code to ensure safety. Local officials use the National Electric Code (NEC) to govern the installation of all wiring and associated hardware such as inverters. "The NEC is in place to protect against shocks and fires caused by electricity," says Mick. "If you ever have a problem that involves an insurance claim and the system was not code compliant when it was installed, it is possible that your insurance company could refuse to honor your claim."

Even if the building department does not require an electrical permit and inspections, they may want to inspect the system if you are connecting to the grid. "It, too, will be looking at whether the system is NEC compliant," notes Mick. "Quite often the utility will require that you submit a single-line diagram of the system designating all electrical components."

In rare instances, Mick has seen local building departments require homeowners to employ a licensed wind system contractor or installer. "No such system of licensing or certifying of wind installers is in place anywhere in the US at this time so no one can meet this requirement." (Small wind installer certification like the one used to certify PV installers is in the works, but it's still several years off.) You should appeal such a requirement on these grounds.

Building department inspectors will visit your site at various stages to ensure that you are doing things correctly. An electrical inspector will also visit at least once to check wiring.

When applying for a building permit, be sure to make it clear that you are planning on installing a small, residential wind machine to offset your own personal electrical demand. Be very clear on this point from the get go. Don't just announce that you want to install a wind turbine on your property. The building department may think you are planning to install a huge commercial wind machine. The unprepared applicant rarely secures a building permit.

"In spite of all potential hurdles, never approach such interactions with a chip on your shoulder," Mick notes. "You can challenge an unfavorable decision or ruling in court, but you will only make enemies in the process, and spend more money that you thought possible." However, do not let anyone deny you the permits you request based on uninformed contentions. Always remember that when it comes to permitting, the burden of proof is on the building department, not you.

"Never go to meetings with a pugnacious attitude or a list of demands either," advises Mick. "From the official's perspective, there is no bigger turnoff than having to deal with an arrogant know-it-all." On the other hand, don't be intimidated by resistance when dealing with building code officials or at public hearings, if any, provided you know what you are talking about. You should expect to be treated in a reasonable and timely manner by everyone involved. Quite often, just letting a building department know that you have consulted with an attorney about your rights as well as your responsibilities (which may be a good idea anyway) will go a long way in reassuring them that you know what you are doing.

Finally, be aware that many building departments have tried to dissuade people from installing wind generators by employing stalling tactics, notes Mick: "To avoid delays, set mutually agreed-upon deadlines for decisions, with homework assignments for each party as appropriate."

And while we are on the subject, avoid the tendency to bypass the law — that is, not obtaining a building or an electrical permit. This can create huge problems. Municipalities have the legal right to force a homeowner to remove his or her wind machine and tower if the homeowner does not secure a required building permit. In fact, we've all heard of municipal governments that have forced homeowners to remove expensive unpermitted wind turbines and towers. Our advice: always obtain all required permits. Do not, under any circumstances, bypass the requirements. The consequences are just too great.

Also, if you are connecting to the grid, be sure to follow the appropriate procedures. Mick knows people who have installed grid-connected systems without permission from the utility and have been denied access to the grid by the local utility company. As a result, they were not able to run their systems.

Obtaining a permit for a wind system may take several months, so apply many months in advance, *and don't buy a wind energy system until your permit has been granted.*

Covenants and Neighbors' Concerns

Covenants from homeowner associations can also create a huge obstacle to wind system installations. Even with permission from the zoning department and a permit from the local building department, restrictive covenants in force in many neighbourhoods can block an installation. Covenants govern many aspects of peoples' lives — from the color of paint they can use on their homes to the installation of privacy fences. They often expressly prohibit renewable energy systems, such as solar hot water systems and solar electric systems, although they may be silent on wind systems, because these systems are rarely installed in neighborhoods.

Unfortunately courts have consistently upheld the legality of covenants imposed by neighborhood associations. If you live in a covenanted community, contact the head of the homeowners or neighborhood association. In Dan's rural mountain neighborhood, the neighborhood association's covenants are overseen by its architectural review committee.

They'll explain the process required to apply for permission to install a wind turbine.

Even in the absence of restrictive covenants, you should consider the needs and desires of your neighbors. "Many people feel strongly about the need to preserve the landscape, views, history and peace and quiet of their neighborhoods," writes Steve Clarke, Canadian wind energy expert. So, unless you live on a large piece of property, be sure to discuss your plans to install a wind turbine with your closest neighbors long before you lay your money down.

Don't expect your neighbors to be as enthusiastic about a wind turbine on your land as you are. Not everyone is enamored of wind energy. Some people seem to have a knee-jerk reaction against wind turbines. Many of them simply fear the unknown. Others seem to object to anything new or environmental. Some neighbors may object out of spite or because of unresolved anger. "If you are feuding with your neighbors, make up with them. They can make your life difficult," advises Robert Preus. Even though your neighbors may have no legal recourse, they can make your life miserable. They could turn up at zoning hearings and voice their opposition, blowing concerns out of proportion.

Although many concerns about wind machines are misguided, be prepared to respond to all concerns without being defensive. You may want to show your neighbors pictures of the wind machine and tower to allay their fears. If you are good at Photoshop, you can even take a photo of your home from their house and place a picture of the wind turbine and tower on your property so they can see what it will look like. If you are thinking about installing a grid-connected system, let them know that you'll be supplying part of their energy, too.

One concern that's terribly misinformed is bird kills. As noted in Chapter 1, most of the concern about bird mortality is from commercial wind farms and is grossly exaggerated. You may want to cite the statistics presented in the first chapter of this book about the perils birds face from domestic cats, windowpanes, automobiles, tall buildings, telecommunications tower and pesticides and how little risk wind machines pose to our fine feathered friends.

Some neighbors may be concerned about interference with radio and television reception. As noted in Chapter 1, the plastic or fiberglass blades of modern residential turbines are transparent to electromagnetic waves such as radio and television. They do not interfere with reception.

Another concern is sound. Although concerns about sound are important, they too are usually blown out of proportion. At a distance of 100 feet, a well-designed, low-rpm residential wind turbine produces a sound level of about 45 decibels. This sound is lower than the background sound level produced in many homes and offices. Most small wind turbines make less sound than a residential air conditioner when operating (not furling). Share this information with your neighbors.

Although turbine sound emissions increase with increasing wind speed so does ambient sound. Background noise on a windy or stormy day such as the rustling of leaves and grasses usually drowns out much of the turbine's sound. There's so much going on that few people can hear a wind machine at all. Because most people are inside in such weather, there's even less chance that a neighbor will hear your wind turbine. You may want to take a neighbor to visit a similar wind turbine to illustrate the point.

While you are at it, be sure to check out local sound ordinances, if any, to ensure you'll be in compliance. Let your neighbors know that you are concerned about protecting them from unwanted sound and are doing everything possible to safeguard them — for example, you are mounting your machine on a very tall tower in a location on your property that minimizes sound beyond your property line. You may want to select a wind machine with a lower tip speed. As noted in Chapter 6, tip speed is the speed of the blade tips. As a rule, the faster the tip speed, the louder the machine.

As a final note on the subject, if you are going to install a freestanding tower be sure to warn neighbors that you'll be using a crane to install it. That way, they won't freak out when they see a monster crane show up for the installation.

To help you allay your neighbors' fears, some companies offer Q & A sheets for residential wind machines. They address many of the common concerns and can be a great resource when warming your neighbors up to the idea of installing a wind turbine on your property. You might make a copy of such material to pass out in person to your neighbors. A typewritten summary of key points may work well. Don't leave material in mailboxes. It's very likely not going to be read, unless you hand it to your neighbors personally. Also, resist the temptation to overwhelm them with written material. It's best to talk to your neighbors in person so you can address concerns directly — before they are blown out of proportion. You may also want to type up a few key talking points and practice them in advance.

Giving neighbors advance notice, answering any questions they have, and being responsive to their concerns is the best way to avoid misunderstandings and problems later, note the folks at Bergey Windpower. "This is doubly good advice if your property size is less than 10 acres or you have to obtain a variance for a building permit. Good neighbor relations boil down to treating your neighbors the same way you would like to be treated and showing respect for their views," they add. "An example of what not to do is to put the turbine on your property line so that it is closer to a neighbor's house than to your own and not give those neighbors any advance notice of your intentions."

Aesthetics

One issue that comes up repeatedly in conversations with neighbors and at zoning hearings for residential wind turbines is aesthetics, a topic that is remarkably difficult to deal with, as Mick notes in an article in *Windletter* (May 2004), adapted here with Mick's permission.

Aesthetics is, of course, highly subjective. Nonetheless, there are steps you can take to make your wind system as visually unobtrusive as possible. As a wind system applicant you can offer some concessions at hearings to assure neighbors that your wind system will not be an eyesore. Mick's list includes the following:

1. The wind system will not be painted a garish color, like hunter orange or electric chartreuse. Wind turbines are painted by the manufacturer, and their colors have been thoroughly considered from two angles: to make sure that they blend in with the environment and to make them distinctive from other manufacturer's turbines. In practice, the first takes precedence over the second. Manufacturers shy away from painting their products in annoying colors.

   Towers are most often made of galvanized steel. They come from the factory bright and shiny, but soon weather to a muted gray color, which blends in with the sky. Several locations across the country require towers to be painted green to blend in with the surrounding vegetation. In almost all the circumstances where Mick has seen this required, the green tower stands out far more than does a weathered-gray, unpainted galvanized tower.

2. The wind system will not display any advertising other than the manufacturer's logo, which is usually on the tail or body of the turbine.

3. The tower will not support any signs, other than perhaps some cautionary signs at the base.

4. The turbine will not be flooded with light at night, as is the case with billboards.

Connecting to the Grid: Working with Your Local Utility

If you are installing a grid-connected system, you'll also need to contact your local utility company to work out an arrangement to connect to their system. You'll need to assure them that the electricity you will be providing will be of the same quality as grid power. Utilities may have questions of procedure, too, notably what happens if their electric lines go down in a storm and they need to send a lineman to work on them. They don't want their linemen getting shocked by a wind system that's operating while their system is down. You'll also need to agree on compensation for any surplus electricity you will be supplying to the grid.

According to Mike Bergey, "After over 200 million hours of interconnected operation, we now know that small utility-interconnected wind turbines are safe, do not interfere with either utility or customer equipment, and do not need any special safety equipment to operate successfully." Moreover, he says, "the output of a wind turbine is made compatible with utility power using either a line-commutated inverter or an induction generator." If you need help on such matters, call your installer or the wind turbine manufacturer. They should be able to help allay the fears and address the concerns of your utility.

As for payment, all utilities in the United States are required by federal law (the Public Utility Regulatory Policy Act of 1978) to buy surplus electricity produced by their grid-connected customers. There's no getting around that. How much they pay is a different matter.

Protecting Your Right to Install a Wind System

Securing permission to install a wind system on a tall tower can be a time-consuming, and sometimes costly ordeal, if you have obstructionist neighbors and a jittery zoning committee. This isn't always the case, however. In Wisconsin and Nevada, for instance, state law limits a zoning department's ability to place unnecessary requirements and burdens on individuals seeking a building permit for a wind system. While many states have solar access laws, Wisconsin and Nevada are the only two that help to protect the right of individuals to install small wind systems.

Federal law requires utilities to pay the "avoided cost" — that is, what it costs the utility to generate the power, which is usually one-fourth to one-third the amount they charge their customers. If you are paying eight cents a kilowatt-hour, the avoided cost may be as low as 2.5 cents. That's all the utility is required to pay.

But that's not the end of the story. Most states have enacted net metering policies that require utilities to pay residential clients the same price they charge. That is, if a utility is charging ten cents a kilowatt-hour, they pay the same rate for power delivered to them. This billing arrangement is known as net metering and is described in Chapter 3.

Some utilities insist on installing two meters, one to track the electricity they sell you, and another to track the electricity you sell them. Check with your local utility to see what their requirements are and how you'll be charged *before* you purchase a system.

Getting Through to the Utility

"Making initial contact with the utility could be one of your most formidable tasks," notes Mick. Unless your local utility has lots of experience with residential wind and solar systems, "in all likelihood, they won't have a specific person who deals with the approval of customer-owned, small scale generation or wind generators," he writes. "Often no one wants to make the crucial decision as to how you can interconnect your equipment to the utility grid."

If your utility is one of the more enlightened ones, they may have a point of contact for renewable energy systems. Check out their website and look for terms such as "net metering," "interconnection agreement," "renewable energy systems," or "distributed generation." You may find all the information you need is available online.

Under federal law, utilities are required to allow customers with wind generators and other renewable energy systems to connect their systems to the utility grid. "While utilities are well aware that they are so required," says Mick, "getting them to cooperate may be another matter entirely." Problems may occur in dealing with rural electric associations, many of which are hostile to the idea of small-scale renewable energy. As a result, it often pays to contact your state's public utility commission (PUC). They regulate utilities in the state. Do so *before* you contact your local utility. The staff at the PUC should be able to describe the approval process

for the various types of utilities (publicly owned, rural electric, and municipally operated) and may also put you in touch with the appropriate contact person at the utility. "When you do finally contact your utility," Mick notes, "it does not hurt to mention that you've been in touch with the public utility commission."

Working Through the Details

"Once in contact with the utility, try to find out who is responsible for processing an interconnection application, and who will be making the final decision on your application," Mick advises. Be prepared to answer all of that person's questions. "If you cannot answer a question, admit it, then find the answer." Remember, your project is not their priority. If they don't know the answer, they're not likely to do the research, especially if they are hostile to the idea of small residential wind.

Before they sign an agreement, your utility may require you to submit blueprints, electrical schematics, and other diagrams that indicate where components will be located. "Its two most important concerns, however, will be an assurance that you have liability insurance and that your equipment includes some sort of system for automatically disconnecting the wind turbine from the grid should there be a power outage," notes Mick. As noted above, all utility-connected inverters for wind systems incorporate some sort of automatic disconnect mechanism should there be a power outage or in the event that the grid goes down. Even so, the utility may require a demonstration, or, at the very least, an inspection.

When dealing with a utility that is inexperienced with grid connections, do not allow them to use their lack of familiarity as an excuse to deny you access, Mick advises. Nor should you allow them to charge you thousands of dollars in engineering fees to review your electrical schematics or test your system. Utility interconnection of wind systems has been occurring for over 30 years and tens of thousands of such systems are connected to the grid across the US, to say nothing of worldwide installations. This is not a new or unproven technology.

Utilities may also require you to install a visible, lockable disconnect — a manually operated switch that allows utility workers to disconnect your wind system from outside your home in case the grid goes down and they need to work on the lines. As you learned in Chapter 2, although grid-connect inverters automatically terminate the flow of electricity to the grid when they detect a drop in line voltage or a change in frequency, the disconnect may be required by your local utility.

Even though lockable disconnects are rarely, if ever, used, do not fight your utility on this requirement, says Mick. "In this and other situations, always remember: utilities often consider 'allowing' you to hook up your wind system to their grid analogous to 'doing you a favor,' even though they are required to by federal law."

Fostering a Productive Relationship

While you may be in the right, keep in mind that the utility is a lot bigger than you are, with much more legal expertise and more resources to fight your installation than you could ever dream imaginable. Your utility can make your approval process very fast or unbelievably difficult. It has all the marbles, and it has the bag, too. And it knows it. Few homeowners have the resources to challenge a utility in court. In a case of irreconcilable differences your best bet is to contact the public utility commission for help.

If you go into the process armed with the proper knowledge, however, it is unlikely ever to come to that, says Mick. It is much better to invite your utility out to inspect your wind system, test the automatic shutdown features, and become a believer. Their staff members will witness for themselves your wind turbine generating electricity and feeding it onto the grid. Mick knows of one homeowner who did just that. He invited the entire utility cooperative out to visit his installation. He ended up hosting the entire company, and he essentially gave a four-hour workshop on wind-generated electricity, answering questions along the way. The upshot? Besides allowing this homeowner to connect his wind system with the grid, the utility became educated on home-sized wind systems and are now advocates of the technology.

Insurance Requirements

Two types of insurance are required before you install your wind system: property damage to protect against *damage to the turbine* and liability insurance to protect you against *damage caused by the turbine*. Both are provided by homeowner's or property owner's insurance policies.

Insuring Against Property Damage

The most cost-effective way to insure a wind system against damage is under an existing homeowner's or property owner's policy. This is far cheaper than trying to secure a separate policy for a wind system.

When you contact your insurance company, Mick advises not to tell them that you want to insure a wind turbine or wind electric generator that will be tied to the utility power grid and back-feed excess generation to the utility. Chances are very good that they will have no idea what you are talking about, and they probably don't have the time or the inclination to learn about this topic on their own. Too much technical information and jargon raises red flags, and up go the rates. Having said that, however, we emphasize that one should never *ever* deceive an insurance company. Besides being illegal, you could find yourself without coverage should you need to file a claim.

Instead, explain that you wish to insure a windmill and tower, both structures that almost everyone is familiar with, as an addition to what is already insured on your property. If you really want to impress the agent, tell her or him that you want to insure an "appurtenant structure" on your current homeowner's policy. This is a term used by the insurance industry to refer to any uninhabited structure on your property not physically attached to your home. Examples include unattached garages, silos, barns, storage sheds, grain elevators, satellite dishes and towers.

Insurance premiums on a homeowner's policy fall into two categories, each with differing rates. Your home is assessed at a higher value than an unattached garage or a storage shed or your tower. This is because people's homes are more lavish than most garages and sheds, and contain a myriad of personal possessions, furniture and clothing not typically found in other structures.

Appurtenant structures, on the other hand, are assessed and charged at a lower rate. It is usual practice for insurance companies to insure appurtenant structures for the total cost of materials plus the labor to build the structure. This represents the installed cost of a wind system.

Coverage of appurtenant structures on a homeowner's policy only applies if the system is installed on the same property as your house. If it is on a separate piece of property, you may have to insure it under a separate policy.

It is best to insure a wind system for its full replacement cost, and not a depreciated value over time. Towers do not wear out, and a properly maintained wind turbine can easily last two decades or more.

Any wind system should have insurance coverage that includes damage to the system from "acts of God" or "acts of Nature," plus possible options for fire, theft, vandalism or flooding. While most wind generator towers are designed to withstand over 120 mile-per-hour winds, tornadoes or hurricanes can wreak havoc on them, just as they would any other structure in their path. Most often, damage is not caused by the extreme wind itself blowing on the tower and turbine, but by debris blown into the tower. Few structures can withstand having a sheet of plywood torn from a garage roof blown into them at 80 miles per hour.

Another "act of Nature" of concern is lightning strikes. A properly installed wind generator tower has a ground rod connected to each tower leg or at each place where a guy cable is connected to the earth or to a concrete anchor, as described in Chapter 6. The wires from the wind generator to the controller and inverter should also be protected by lightning arrestors. Grid-connected systems are also grounded on the utility side of the inverter. Adding a lightning arrestor and voltage surge arrestors on the utility side of the inverter affords additional protection. While none of this will guarantee that your system will not be damaged by a lightning strike, it certainly reduces its likelihood. Plus, in the eyes of the insurance company, you have taken prudent measures to protect your system.

Fire is of minimal concern to a wind electric system. However, it should go without saying that the wiring of the entire system must comply with the electric code. If you have a fire in the house caused by some funky wiring in the wind system, your claim may be denied and your policy canceled.

Theft of an entire wind system, or even any part of the system, seems unlikely, so should not be a major issue. Vandalism, on the other hand, may be a bigger concern for wind turbine owners. While incidents of vandalism are infrequent, they have occurred. The most frequently filed claim attributed to vandalism involves guns being fired at the turbine's blades or the generator itself. In either case, damage can be substantial.

Flood insurance is a nationally administered program that is usually geared to damage to primary dwellings. Costs can be exceptionally high for a home located along a coastline or in a floodplain near a stream or river with a penchant for flooding. Since this is a very site-specific assessment, no insurance company Mick has contacted would even quote a range of prices. If you live in a floodplain, get an estimate before beginning construction.

Insuring a wind system as an appurtenant structure on a homeowner's policy is relatively inexpensive. While a percentage of the home insurance coverage extends to appurtenant structures, added insurance can be purchased. In the Midwest where Mick lives, annual premiums for additional coverage are about $2.50 per $1,000 of value (as of May 2008). Since most rural homeowner insurance claims are for fire damage, the deciding factor in pricing coverage is determined by the homeowner's distance from the nearest fire department, regardless of whether your wind turbine can catch fire or not. As a result, the additional premium would likely be slightly lower for a system in town, and slightly higher for a system sited in the country.

Liability Coverage

Homeowner's insurance protects against damage *to* the turbine. Liability insurance also part of a homeowners policy protects you, the owner, against damage caused *by* the turbine, for example, if the tower falls on a neighbor's garage. It also protects you from personal injury or death of employees working on a utility line during a power outage caused by your system. Even though this can't occur because of the protective mechanisms built into inverters, utilities insist on this coverage.

Liability coverage is relatively inexpensive; and, as just noted, it comes with a homeowner's policy. In most locales, liability coverage for your home is $300,000. This is the minimum coverage required for anyone with a federally insured mortgage. Increasing this to $500,000 coverage may add an additional $10 to an annual premium in most areas. Extending coverage to one million may add $35 to $40 more to the annual premium. It is advisable to ask about an umbrella policy for liability coverage of $1,000,000 or more, as the rates are cheaper still.

If your wind turbine is a grid-connected system, the local utility may dictate the level of liability insurance that it requires as a condition for interconnection. If the amount seems unreasonable to you, file an appeal with your state's public utility commission. If you live in a rural area, ask the utility what liability insurance requirements they have for farmers with emergency standby generators. Since these systems have similar protective devices, the requirements should be similar. By doing this, you can assure yourself that the utility is not merely upping the ante for your wind system to dissuade you from competing with them by growing your own electricity.

Buying a Wind Energy System

If wind energy seems like a good financial, social, or recreational pursuit for you, we strongly recommend hiring a competent, experienced professional wind site assessor and a professional installer. A local supplier/installer with experience and knowledge can be a great ally. They will supply all of the equipment, be certain that it is compatible, help obtain permits, if necessary, and install the system. They'll test the system to be sure it is operating satisfactorily and will be there to answer questions and to address any problems you have with the equipment.

We recommend that you find an installer who also provides routine maintenance and one who stands solidly by his or her work. Most wind turbines are guaranteed for five years, but you want to be sure the installer will respond promptly and knows what he or she is doing. Look for people who've been in the business for a while — the longer, the better — and who have installed a lot of systems. Local installers who know the business are worth their weight in gold.

As in any major home project, ask for references, and call them. Visit wind installations, if possible, and contact the local office of the Better Business Bureau. Get everything in writing. Sign a contract. Be sure the installer has insurance to protect employees during installation. Don't pay for the entire installation up front. Be sure you are on the site when the work is done. The box includes a list of questions to ask potential installers.

You can also purchase equipment from a local or online supplier, and put it up yourself with or without their guidance. We don't recommend this route unless you are handy and you have attended a couple of workshops.

Raising a tower is risky business. Wiring is fraught with difficulties. Connecting to the electrical grid is a job for professionals. A wind energy system is such a huge investment that you don't want to mess it up.

Another option is to buy from a local supplier or an online source and sponsor a workshop on your property to take care of the installation. Nonprofit organizations like Solar Energy International and the Midwest Renewable Energy Association are often looking for wind energy installations in different parts of the country. They fly in wind energy experts like Mick or Ian who teach a one- or two-week installation workshop. Workshop registrants pay a fee to cover the cost of the instructor, the costs of advertising and setting up the workshop, and perhaps food, drinks, snacks and even a toilet. Their fees do not underwrite materials costs. The homeowner pays the costs of materials. A local contractor oversees the preparation and installation. (He'll be the installer of record, required for the permit.)

Although you may not save any money on the installation, workshop installations can be very satisfying. You do have to be comfortable with a dozen or more people on your site for a week. If the workshop leader is competent and checks all of the attendees' work, you'll get a wind machine installed while helping a group of people learn about wind electricity. Be sure to contact these nonprofit organizations well in advance and be prepared to help organize and publicize the workshop. Also, be sure your insurance will cover volunteer workers on your site.

Parting Thoughts

You have now reached the end of this book. When you started reading you may have simply wanted to determine if a wind system was suitable for your home or business. Perhaps you were sure you wanted to install a wind system, but didn't know how to proceed. We hope that you now have a clear understanding of what is involved.

So, what will you do with this new knowledge?

If you have come to realize that your dreams for a wind system were unrealistic, you can pursue other dreams — and be glad that you did not spend a lot of money on a wind system that would not have met your expectations. If, however, you now know that a wind system is right for you, perhaps the next step is to investigate available systems, find a dealer/installer, and start harnessing the wind.

With ongoing inflation and increasing demand for the materials that go into a wind system — steel, copper, cement, etc. — prices of wind turbines and towers are not likely to come down. The longer you wait, the higher the cost of installing a wind system is likely to be. The cost of electricity and the fuels needed to produce it are also rising. The sooner you get your wind system up and running, the sooner you will start saving. Time's a-wasting.

But our job is over. We've helped you get up to speed on wind energy systems. You now know a great deal about wind, wind energy systems, electricity, wind turbines, towers, inverters, batteries and the maintenance requirements of wind energy systems. You've obtained a good solid working knowledge — good theoretical as well as practical information — that puts you in a good position to move forward. We've given you mountains of advice on installation and helped you grapple with economic issues. Our work has ended, but yours is just beginning.

We wish you the best of luck! May the wind be as much an ally to you as it has been to us.

Questions to Ask Potential Installers

1. How long have you been in the business? (The longer the better)

2. How many systems have you installed? How many systems like mine have you installed? (The more systems the better).

3. How will you size my system?

4. Do you provide recommendations to make my home more energy efficient first? (As stressed in the text, energy efficiency measures reduce system size and can save you a fortune.)

5. Do you carry liability and worker's compensation insurance? Can I have the policy numbers and name of the insurance agents? (Liability insurance protects against damage to your property. Worker's compensation insurance protects you from injury claims by the installer's workers.)

6. Are you bonded? (Bonding, as explained in the text, provides homeowners with financial recourse if an installer does not meet his or her contractual obligations.)

7. What additional training have you undergone? When?

8. Will employees be working on the system? What training have they received? How many systems have they installed? Will you be working with the crew or overseeing their work? If you're overseeing the work, how often will you check up on them? If I have problems with any of your workers, will you take care of them immediately? (Be sure that the owner of the company will be actively involved in your system or that he or she sends an experienced crew to your site.)

9. Are you a licensed electrician or will a licensed electrician be working on the crew? (State regulations on who can install a wind system vary. A licensed electrician is usually not required, except to pull the permit, supervise the project, and make the final connection to your electrical panel.)

10. What brand turbines and inverters will you use? Do you install UL listed components? (To meet code, all components must be UL listed or listed by some other similar organization.)

11. Do you guarantee your work? For how long? What does your guarantee cover? How quickly will you respond if troubles emerge? (You want an installer who guarantees the installation for a reasonable time and who will fix any problems that arise immediately.)

12. Do you offer service contracts? (Service contracts may be helpful early on to be sure ☛

the system runs flawlessly.) How much will a service contract cost? What does it cover?

13. Can I have a list of your last five projects with contact information? (Be sure to call references and talk with homeowners to see how well the installer performed and how easy he or she was to work with.)

14. What is the payment schedule? Can I withhold the final 10% of the payment for a week or two to be sure the system is operating correctly? (Don't pay for a system all at once. A deposit, followed by one or two payments protects you from being ripped off. Never make a final payment until you are certain the system is working fine.)

15. Will you pull and pay for the permits?

16. Will you work with the utility to secure an interconnection agreement?

17. What's a realistic schedule? When can you obtain the equipment? When can you start work? How long will it take? ■

Endnotes

Chapter 1

1. Studies show that bat mortality usually occurs in the late summer and fall when bats are migrating or dispersing. Researchers believe that bat deaths occur at this time because the bats may switch off their echolocation to conserve energy while migrating.

2. Ice buildup increases the drag in relation to the aerodynamic lift, which dramatically slows the blades.

Chapter 2

1. Humid air is less dense because water (H_2O) has an atomic weight of 18. The main components of air, nitrogen (N_2) and oxygen (O_2), have molecular weights of 28 and 32, respectively. So, as water vapor displaces nitrogen and oxygen molecules, the air gets less dense.

2. Manufacturers routinely report rotor diameter. The rotor diameter is the diameter of the circle described by the spinning blades and is often greater than twice the "blade length." That's because the blades attach to the hub.

3. Ian's not convinced that any power rating will be a good measure.

Chapter 3

1. Winchargers and other early battery-charging systems used DC generators, which produced DC electricity. Some modern battery-charging wind turbines incorporate AC generators but rectify (convert) the AC to DC at the generator and send DC down the tower. Examples are the Whisper 100 and Bergey XL.1.

2. In freestanding towers, these functions are provided by the tower itself, and its

foundation, so both the tower and the foundation must be much stronger.

3. Because the term "grid" refers to the high-voltage transmission system, not local electrical distribution systems, customers are connected to the grid through their local utility and indirectly to its distribution system, the grid. Because of this, the terms "grid-connected" and "grid-tied," which are used by the renewable energy community, are not entirely accurate. "Utility-connected" or "utility-tied" would be better terms, but that is not the common usage in the renewable energy world.

4. Although most small wind turbines produce wild AC, commercial-size wind turbines and one small wind turbine, the Endurance, which are induction machines, produce grid-compatible AC electricity.

5. In the state of Washington, grid-connected systems with battery back have three meters if the owner wants to apply for a production incentive. (Financial incentive based on electrical production of the system.) This requirement prevents clients from buying energy at lower rates to charge batteries and then selling electricity from their batteries at higher "solar" rates.

Chapter 4

1. State wind maps contain estimates, not direct measurements of wind speed. Direct measurements are used to validate the maps, but maps are created using upper air wind speed data (wind balloon data) and topographic data.

2. As long as both height numbers are in consistent units (meters or feet) and both speed numbers are in consistent units (meters per second or miles per hour), the results will be OK, even if height is in feet and speed is in meters per second.

Chapter 5

1. As noted in Chapter 3, the blades of a turbine attach to a face plate and therefore do not meet in the center of the rotor. As a result, the radius is often a little greater than the blade length.

2. Some vertical furling wind turbines like the Aeromax Lakota have a spring that complements gravity return.

3. In airplanes with variable pitch propellers, the engine and propeller are operated at a relatively constant speed and power is controlled by varying pitch of the propellers.

4. An axial flux alternator consists of a round plate that contains the permanent magnets. It is attached perpendicularly to the rotor.

Chapter 6

1. Steel and concrete prices are increasing rapidly due to the rising cost of energy, so these prices will not reflect current prices. The important thing to focus on is the cost of one tower or foundation relative to another.

2. Some individuals recommend the use of a device known as a come along. Unfortunately, a come along has a limited length of pull and can drop its load unexpectedly, so it is not safe for lifting or hoisting.

Chapter 7

1. The amp-hour capacity is based on a specified discharge rate, typically 8, 10 or 20 hours. If a battery is discharged at the specified rate, it should yield the specified, or rated, amp-hours. If it is discharged at a higher rate, it will yield fewer amp-hours than rated. If it is discharged at a slower than specified rate, it will yield more than its rated amp-hours.

Chapter 8

1. Each coil has multiple turns and each coil is known as a winding. The voltage step-up or step-down ratio is proportional to the turns ratio.

2. Power factor is a measure of how well the voltage and the current waveforms are aligned. Inductive loads such as motors cause the current to lag behind the voltage. If, for example, the current lags behind the voltage by 90 degrees (one quarter cycle), they are so far out of alignment that no real power can be delivered, although the current remains the same. Current flowing in wires causes voltage drop, which lowers the voltage at the end of the line. A factory with a "poor power factor" requires far higher current than is needed to power loads with a good power factor. The power factor is a number between zero and one.

3. One exception is the Skystream, the output of which can be connected to the AC-side of the inverter, so it does not charge the batteries directly.

Chapter 10

1. Some states, like Wisconsin and Colorado, have state laws that supersede local zoning authority, allowing wind turbines to be installed on tall towers for optimum performance.

Resource Guide

To view a complete list of resources, see my website: danchiras.com.

Books

Bartmann, Dan and Dan Fink. *Homebrew Wind Power: Build Your Own Wind Turbine: A Hands-On Guide to Harnessing the Wind.* Buckville Publications, LLC, 2008. A clear and comprehensive guide to building a quiet, efficient, reliable and affordable wind turbine.

Chiras, Dan. *The Homeowner's Guide to Renewable Energy.* New Society Publishers, 2006. Contains information on residential wind energy and solar electric systems.

Ewing, Rex A. *Power With Nature: Solar and Wind Energy Demystified.* PixyJack Press, 2003. Skip to the second half if you want to get to the meat of the matter.

Gipe, Paul. *Wind Power: Renewable Energy for Home, Farm, and Business.* Chelsea Green, 2004. Somewhat technical introduction to small and large (commercial) wind generators.

Gipe, Paul. *Wind Energy Basics: A Guide to Small and Micro Wind Systems.* Chelsea Green, 1999. A brief and somewhat technical introduction to wind energy for newcomers.

Manwell, J. F., J. G. McGowan, and A.L. Rogers. *Wind Energy Explained.* John Wiley and Sons, 2002. A comprehensive textbook that focuses on commercial wind energy.

National Renewable Energy Lab. *Wind Energy Information Guide.* University Press of the Pacific, 2005. A brief introductory pamphlet on wind energy.

Piggott, Hugh. *Wind Power Workshop.* Center for Alternative Technology, 1997. A guide for those who want to build their own wind turbines and towers by Europe's small wind expert.

Rohatgi, Janardan S. and Vaughn Nelson. *Wind Characteristics: An Analysis for the Generation of Wind Power.* Alternative Energy Institute, 1994. The best book available on siting wind turbines. Contains lots of math. You can obtain a copy from the American Wind Energy Association or Alternative Energy Institute at (806) 651-2295.

Scheckel, Paul. *The Home Energy Diet.* New Society Publishers, 2005. A great guide for energy conservation in homes.

Wegley, H. L., James V. Ramsdell, Montie M. Orgill, and Ron L. Drake. *A Siting Handbook for Small Wind Energy Conversion Systems.* WindBooks, 1980. Out-of-print guide to siting a wind turbine.

Articles

Brown, Lester R. "Harnessing the Wind for Energy Independence," *Solar Today* 16 (2), 24–27, 2002. Good overview of the prospects of commercial wind energy.

Green, Jim. "Small Wind Installations in Colorado," *Solar Today* 14 (1), 20–23, 2000. Several case studies of wind installations.

Green, Jim and Mick Sagrillo. "Zoning for Distributed Wind Power — Breaking Down the Barriers." Conference Paper NREL/CP-500-38167, May 2005. National Renewable Energy Association, Golden, Colorado.

Gipe, Paul. "Small Wind Systems Boom," *Solar Today* 1 (2), 29–31, 2002. Good overview of small wind energy systems by one of America's leading experts on the subject.

Dankoff, Windy. "Lightning Happens," *Home Power* (107), 60–64. Important information on protecting a renewable energy system from lightning.

Dankoff, Windy. "Top Ten Battery Blunders," *Home Power* (115), Aug./Sept. 2006, 54–60. Important reading for anyone installing a battery bank.

Home Power Staff. "Clearing the Air: Home Power Dispels the Top RE Myths," *Home Power* 100, 32–38, 2004. Superb piece for those who want to learn about the many false assumptions and beliefs regarding renewable energy.

Hackleman, M. and C. Anderson. "Harvest the Wind," *Mother Earth News* 192, 70–78, 2002. An easy-to-read and fact-filled article on wind power.

Kuebeck, Sr., Peter and Peter Kuebeck, Jr. "Old Jacobs — Current Again," *Home Power* 89, 70–78, 2002. Personal story about a great old wind turbine that's being refurbished and installed throughout the country.

Laughlin, Don. "You Gotta Have Height: A Tower Construction Project in Iowa," *Home Power* 92, 30–38, 2003. Interesting case study. Well worth reading to learn more about tower construction and erection.

Moore, Kevin. "Pumping Water with the Wind," *Home Power* (122), Dec./Jan. 2008, 88–93. A good overview of water-pumping wind-mills.

Osborn, D. "Winds of Change." *Solar Today* 17 (6). 22–25, 2003. Looks at a number of important issues, include bird mortality by wind towers compared to other sources.

Pahl, Greg. "Choosing a Backup Generator," *Mother Earth News* 202, 38–43, 2004. A fairly detailed overview of what to look for when buying a backup electrical generator for a RE system.

Pearen, Craig. "Brushless Alternators," *Home Power* 97, 68–71, 2003. Brief but interesting look at low-maintenance brushless alterna-tors for wind machines.

Piggott, Hugh. "Estimating Wind Energy," *Home Power* 102, 42–44, 2004. A brief piece that will help you determine how much elec-trical energy you can produce on your site.

Perez, Richard. "Flooded Lead-Acid Battery Maintenance," *Home Power* 98, 76–79, 2004.

Read, memorize this, and put its advice into practice if you plan on installing a stand-alone RE system.

Preus, Robert. "Thoughts on VAWTs: Vertical Axis Wind Generator Perspectives," *Home Power* 104, 98–100, 2005. Excellent article with excellent photos and diagrams.

Russell, Scott. "Starting Smart: Calculating Your Energy Appetite," *Home Power* 102, 70–74, 2004. Great introduction to using household load analysis to determine your household electrical demand.

Sagrillo, Mick. "Aesthetic Issues and Residential Wind Turbines," *Windletter* 23 (5), 1–2. A must-read for anyone who is interested in installing a wind turbine on his or her property. Many more of Mick's articles on small wind from *Windletter* are available at awea.org/smallwind/sagrillo

_____. "An Open Letter: To Inventors of Vertical Axis Wind Turbines and Rooftop Wind 'Technological Breakthroughs,'" *Windletter* 27 (3), 2008, 1–6. A frank discussion of vertical axis wind turbines.

_____. "Annual Wind Turbine Inspections," *Windletter* 21 (5), 2002, 1–3. Good advice for anyone who installs a wind turbine. Many of Mick's articles on small wind from *Windletter* are available at awea.org/ small-wind/sagrillo/index.html

_____. "Bats and Wind Turbines," *Windletter* 22 (2), 2003, 1–2. Important reading for anyone concerned about this issue.

_____. "Buying Used Wind Equipment," *Windletter* 21 (11), 2002, 1–3. Sound advice on buying used wind turbines.

_____. "The Wind's Power at Various Tower Heights," *Windletter* 24 (12), 2005, 1–2. A compelling look at wind speed vs. tower height and turbine output.

_____. "Incremental Tower Costs versus Incremental Energy," *Windletter* 25 (1), 2006, 1–3. Looks at the cost of taller towers and the additional return on this investment.

_____. "Payback: The Wrong Question," *Windletter* 26 (7), 2007, 1–3. Examines the topic of payback, especially why it is the wrong way to think about a renewable energy system.

_____. "Planning Your Wind System (2) — Building Permits," *Windletter* 25 (7), 2006, 1–4. Will help you understand building permits and what you'll need to consider when applying for a permit.

_____. "Planning Your Wind System — Evaluating Your Wind Resource," *Windletter* 26 (1), 2007, 1–3. Examines issues such as load assessment, wind speed monitoring and wind maps.

_____. "Planning Your Wind System — Homeowner's Insurance," *Windletter* 25 (11), 2006, 1–3. What every homeowner interested in installing a wind system needs to know about homeowner insurance to cover cost of damage to a wind turbine.

_____. "Planning Your Wind System — Liability Insurance," *Windletter* 25 (10), 2006, 1–2. What every homeowner interested in installing a wind system needs to know about liability insurance.

_____. "Planning Your Wind System — Utility Requirements," *Windletter* 25 (8), 2006, 1–3. Sound advice on working with the utility when installing a grid-connected wind system.

_____. "Planning Your Wind System — Working with Your Neighbors," *Windletter* 27 (1), 2008, 1–4. Read this before announcing to your neighbors that you're going to install a wind turbine — in fact, before you buy your turbine and tower.

_____. "Residential Wind Turbines and Lightning," *Windletter* 22 (11), 2003, 1–3. Describes lightning protection measures for small wind turbines.

_____. "Residential Wind Turbines and Noise," *Windletter* 23 (3), 2004, 1–4. Important information on sound levels and allaying concerns of neighbors.

_____. "Rules of Thumb for Siting Wind Turbines," *Windletter* 25 (6), 2006, 1–3. Good description of things you need to know when determining tower height for a wind energy system.

_____. "Tall Tower Economics," *Windletter* 25 (2), 2006, 1–4. Discusses the incremental increase in energy output of turbines compared to the incremental cost of installing taller towers.

_____. "The Wind's Power at Various Tower Heights," *Windletter* 24 (12), 2005, 1–3. A great discussion of tower heights that helps beginners realize the importance of tower height when installing a wind machine.

Sagrillo, Mick and Ian Woofenden. "Wind Turbine Buyer's Guide," *Home Power* 119, 2007, 34–40. A brief overview of wind turbines and details on several popular wind turbines.

Schwartz, Joe. "What's Going On — The Grid?" *Home Power* (106), 2005, 26–32. Excellent look at many of the grid-tied inverters.

Short, Walter and Nate Blair. "The Long-Term Potential of Wind Power in the US," *Solar Today* 17 (6): 28–29, 2003. Important study of wind power's potential.

Sindelar, Allan and Phil Campbell-Graves. "How to Finance Your Renewable Energy Home," *Home Power* 103, 2004, 94–99. Very useful article.

Stone, Lauri. "Hiring a PV Pro," *Home Power* 114, 2006, 48–51. Although this article is about hiring a professional installer for a PV system, much of what the author says applies to wind system installers.

Swezey, Blair and Lori Bird. "Buying Green Power — You Really Can Make a Difference," *Solar Today* 17(1), 2003, 28–31. An in-depth look at ways (such as green tags) homeowners can tap into renewable energy without installing a system on their home.

Woofenden, Ian. "Wind Generator Tower Basics," *Home Power* (105), 2005, 64–68. Excellent overview of tower types.

Woofenden, Ian. "Battery Filling Systems of the Americas: Single-Point Watering System," *Home Power* 100, 2004, 82–84. This article is a must for those who would like to reduce battery maintenance.

Woofenden, Ian and Mick Sagrillo. "How to Buy a Wind-Electric System," *Home Power* 122, 28–34. Terrific review of the steps you need to take when contemplating a wind turbine plus a great overview of 25 residential wind turbines.

Magazines

BackHome. P.O. Box 70, Hendersonville, NC 28793. Tel: (800) 992-2546. Website: BackHomeMagazine.com. Publishes articles on renewable energy and many other subjects for those interested in creating a more self-sufficient lifestyle.

Backwoods Home Magazine. P.O. Box 712, Gold Beach, OR 97444. Tel: (800) 835-2418. Website: backwoodshome.com. Publishes articles on all aspects of self-reliant living, including renewable energy strategies.

The CADDET Renewable Energy Newsletter. 168 Harwell, Oxfordshire OX11 ORA, United Kingdom. Tel: +44 123335 432968. Website: caddet.org. Quarterly magazine published by the CADDET Centre for Renewable

Energy. Covers a wide range of renewable energy topics.

EREN Network News. Newsletter of the Department of Energy's Energy-Efficiency and Renewable Energy Network. Website: apps1.eere.energy.gov/news/

Home Energy Magazine. 2124 Kittredge Street, No. 95, Berkeley, CA 94704. Website: homeenergy.org. Great resource for those who want to learn more about ways to save energy in conventional home construction.

Home Power. P.O. Box 520, Ashland, OR 97520. Tel: (800) 707-6585. Website: homepower. com. Publishes numerous extremely valuable how-to and general articles on renewable energy, including solar hot water, PVs, wind energy, microhydroelectric and occasionally an article or two on passive solar heating and cooling. This magazine is a goldmine of information, an absolute must for anyone interested in learning more. It also contains important product reviews and ads for companies and professional installers. They also sell CDs containing back issues.

Mother Earth News, 1503 SW 42nd St., Topeka, KS 66609. Website: motherearthnews.com. One of my favorite magazines. Usually publishes a very useful article in each issue on some aspect of renewable energy.

Solar Today. ASES, 2400 Central Ave., Suite G-1, Boulder, CO 80301. Tel: (303) 443-3130. Website: solartoday.org. This magazine, published by the American Solar Energy Society, contains lots of good information on passive solar, solar thermal, photovoltaics, hydrogen, and other topics, but not much how-to information. Also lists names of engineers, builders and installers and lists workshops and conferences. Mick Sagrillo writes a monthly column for them.

Organizations

American Wind Energy Association. 122C Street, NW, Suite 380, Washington, D.C. 20001. Tel: (202) 383-2500. This organization sponsors an annual conference on wind energy. Check out their website which contains a list of publications, their online newsletter, frequently asked questions, news releases and links to companies and organizations.

British Wind Energy Association. 26 Spring Street, London W2 1JA. Tel: 0171 402 7102. Website: bwea.com. Actively promotes wind energy in Great Britain. Check out their website for fact sheets, answers to frequently asked questions, links and a directory of companies.

Center for Alternative Technology. Machynlleth, Powys SY20 9Az. Tel: 01654 703409. Website: cat.org.uk. This educational group in the United Kingdom offers workshops on alternative energy, including wind, solar, and microhydroelectric.

Center for Renewable Energy and Sustainable Technologies. Holywell Park, Loughborough University, Loughborough, Leicestershire, LE11 3TU. Website: lboro.ac.uk/crest. Active in research and education in renewable energy.

European Wind Energy Association. Rue du Trone 26, B-1000, Brussels, Belgium. Tel: +32 2 546 1940. Website: ewea.org. Promotes wind energy in Europe. The organization publishes the *European Wind Energy Association Magazine*. Their website contains information on wind energy in Europe and offers a list of publications and links to other sites.

Energy Efficient Building Association. 490 Concordia Ave., P.O. Box 22307, Eagen,

MN 55122. Tel: (651) 268-7585. Website: eeba.org. Offers conferences, workshops, publications and an online bookstore.

National Wind Technology Center of The National Renewable Energy Laboratory. 1617 Cole Blvd., Golden, CO 80401-3393. Tel: (303) 275-3000. Website: nrel.gov/wind. Their website provides a great deal of information on wind energy, including a wind resource database.

Solar Energy International. P.O. Box 715, Carbondale, CO 81623. Tel: (970) 963-8855. Website: solarenergy.org. Offers a wide range of workshops on solar energy, wind energy and natural building.

Solar Living Institute. P.O. Box 836, Hopland, Ca 95449. Tel: (707) 744-2017. Website: solarliving.org. A nonprofit organization that offers frequent hands-on workshops on solar energy and many other topics. Be sure to tour their facility if you are in the neighborhood.

US Department of Energy and Environmental Protection Agency's ENERGY STAR program. Tel: (888) 782-7937. Website: energystar.gov.

US Department of Energy's Energy Efficiency and Renewable Energy. Website: eere.energy.gov.

Small Wind Turbine Manufacturers

Abundant Renewable Energy, 22700 NE Mountain Top Road, Newberg, OR 97132. Tel: (503) 538-8298. Website: abundantre.com.

Aerostar, Inc., PO Box 52, Westport Point, MA 02791. Tel: (508) 636-5200. Website: aerostarwind.com.

Bergey Windpower, 2200 Industrial Blvd., Norman, OK 73069. Tel: (405) 364-4212. Website: bergey.com.

Bornay Wind Turbines, P.I. RIU, Cno. Del Campanar, 03420 Catalla (Alicante) Spain. Tel: (965) 560-025 (966) 543-077. Website: bornay.com.

Flowtrack Australia, End of Turntable Falls Road, Nimbin, Australia. Tel: 0266 891431. Website: flowtrack.com.au.

Iskra Wind Turbines, Ltd. The Innovation Centre, Epinal Way, Loughborough LE11 3EH, United Kingdom. Tel: 0845 8380588. Website: iskrawind.com.

Jacobs Wind Turbines, Wind Turbine Industries Corp., 16801 Industrial Circle S.E., Prior Lake, Minnesota 55272. Tel: (952) 447-6064. Website: windturbine.net.

Kestrel Wind Turbines, P.O. Box 3191, North End, Port Elizabeth 6056, South Africa. Tel: +27 (0)41 401 2645. Website: kestrelwind.co.za.

Marlec Engineering Co. Ltd., Rutland House, Tevithick Road, Corby, Northants NN17 5XY, United Kingdom. Tel: +44 (0)1535 201588. Website: marlec.co.uk

Proven Energy Ltd., Wardhead Park, Stewarton, Ayrshire, KA3 5LH, Scotland, UK. Tel: 0044 (0) 1560 485 570. Website: provenenergy.co.uk.

Southwest Windpower, 1801 W. Route 66, Flagstaff, AZ 86001. Tel: (928) 779-9463. Website: windenergy.com.

West Wind, J.A. Graham Group, 3 Carmavy Road, Crumlin, Co. Antrim, BT29 4TF, Northern Ireland. Tel: 0044 (0) 28 9445 2437. Website: westwindturbines.co.uk.

Wind Turbine Manufacturers

Abundant Renewable Energy
    Website: abundantre.com
Bergey Windpower Website: bergey.com

Energy Maintenance Systems
    Website: energyms.com
Endurance Wind Power
    Website: endurancewindpower.com
Entegrity Wind Systems
    Website: entegritywind.com
Pine Ride Products (Eoltech)
    Website: pineridgeproducts.com
Halus Power Systems Website: halus.com
Kestrel (Imported by DC Power Systems)
    Website: dcpower.com
Energie PGE Website: energiepge.com
Proven (Imported by Alaska RE) (remotepow-
    erinc.com) Lake Michigan Wind ans Sun
    (windandsun.com) and Solar Wind
    Works
    (solarwindworks.com)
Southwest Windpower
    Website: windenergy.com
True-North Power NG, Inc.
    Website: truenorthpower.com
Ventera Website: venteraenergy.com
Wind Turbine Industries Corp
    Website: windturbine.net

Inverter Manufacturers

Beacon Power Corp
    Website: beaconpower.com
Fronius USA Website: fronius.com
Magnetek, Inc.
    Website: alternative-energies.com
Outback Power Systems
    Website: outbackpower.com
PV Powered LLC Website: pvpowered.com
Sharp Electronics
    Website: sharp-usa.com/solar
SMA America, Inc.
    Website: sma-america.com
Xantrex Technology, Inc.
    Website: xantrex.com

Charge Controllers

Apollo Solar Website: apollo-solar.net
Blue Sky Energy, Inc.
    Website: blueskyenergyinc.com
MidNite Solar, Inc.
    Website: midnitesolar.com
Morningstar Corp.
    Website: morningstarcorp.com
OutBack Power Systems
    Website: outbackpower.com
Xantrex Technology, Inc.
    Website: xantrex.com

Battery Manufacturers

Flooded Lead Acid
Deka/MK Website: eastpenndeka.com
Exide Battery
    Website: hawkerpowersource.com
GB HUP Website: enersysmp.com
Power Batter
    Website: powerbattery.com
Surrette/Rolls Battery
    Website: surrett.com
Trojan Battery
    Website: trojan-battery.com
US Battery Website: usbattery.com

Sealed Batteries

Concorde Battery
    Website: concordebattery.com
Deka/MK
    Website: eastpenndeka.com
Discover Energy
    Website: discoverenergy.com
Exide Website: exide.com
Exide Sonnenschein Website: exide.com
FullRiver Website: fullriver.com
Hawker Website: hawkerpowersource.com
Power Battery Website: powerbattery.com
Trojan Website: trojan-battery.com

Battery Monitor Manufacturers

Bogart Engineering
    Website: bogartengineering.com
OutBack Power Systems
    Website: outback.com
Xantrex Website: xantrex.com

Wind Turbine Tower Kits

Abundant Renewable Energy, 22700 NE Mountain Top Road, Newberg, OR 97132. Tel: (503) 538-8298.
Website: abundantre.com.

Bergey Windpower, 2200 Industrial Blvd., Norman, OK 73069. Tel: (405) 364-4212. Website: bergey.com.

IDC Solar, Wind & Water, P.O. Box 630, Chino Valley, AZ 86323. Tel: (928) 636-9864. Website: idcsolar.com.

Independent Power Systems, 1501 Lee Hill Road #19, Boulder, CO 80304. Tel: (303) 443-0115. Website: solarips.com.

Lake Michigan Wind and Sun, 1015 County Road U, Sturgeon Bay, WI 54235. Tel: (970) 743-0456. Website: windandsun.com.

Otherpower.com, 2606 W. Vine Drive, Fort Collings, CO 80521. Tel: (877) 944-6247. Website: otherpower.com.

Southwest Windpower, 1801 W. Route 66, Flagstaff, AZ 86001. Tel: (928) 779-9463. Website: windenergy.com.

The Energy Depot, Inc. 16650 Jane Street, Kettleby ON LOG 1J0 Canada. Tel (905) 760-7511. Website: energydepot.ca.

Index

About the Authors

Dan Chiras is an internationally acclaimed author who has published over 24 books, including *The Homeowner's Guide to Renewable Energy* and *Green Home Improvement*. He is a certified wind site assessor and has installed several residential wind systems. Dan is director of The Evergreen Institute's Center for Renewable Energy and Green Building (www.evergreeninstitute.org) in east-central Missouri where teaches workshops on small wind energy systems, solar electricity, passive solar design, and green building. Dan also has an active consulting business, Sustainable Systems Design (www.danchiras.com) and has consulted on numerous projects in North America and Central America in the past ten years. Dan lives in a passive solar home powered by wind and solar electricity in Evergreen, Colorado.

Mick Sagrillo is a world-renowned authority on small wind energy and has been in the business since 1981. Mick teaches workshops on wind energy through the Midwest Renewable Energy Association, Solar Energy International, and other organizations. Mick has published numerous articles on all aspects of small wind for *Home Power, Solar Today,* and the *American Wind Energy Association's Windletter*. Mick served as the wind technology specialist for Wisconsin's Focus on Energy.

Dan Chiras.

Ian Woofenden is a senior editor for *Home Power* magazine and has written a number of excellent articles on small wind energy for the magazine. Ian teaches workshops and is a workshop coordinator. He and his family have lived on wind and solar energy for many years in Washington State's San Juan Islands.

If you have enjoyed *Power from the Wind* you might also enjoy other

BOOKS TO BUILD A NEW SOCIETY

Our books provide positive solutions for people who want to
make a difference. We specialize in:

Sustainable Living • Green Building • Peak Oil • Renewable Energy
Environment & Economy • Natural Building & Appropriate Technology
Progressive Leadership • Resistance and Community
Educational and Parenting Resources

New Society Publishers

ENVIRONMENTAL BENEFITS STATEMENT

New Society Publishers has chosen to produce this book on Enviro 100, recycled paper
made with **100% post consumer waste**, processed chlorine free, and old growth
free.

For every 5,000 books printed, New Society saves the following resources:[1]

37	Trees
3,369	Pounds of Solid Waste
3,706	Gallons of Water
4,834	Kilowatt Hours of Electricity
6,124	Pounds of Greenhouse Gases
26	Pounds of HAPs, VOCs, and AOX Combined
9	Cubic Yards of Landfill Space

[1]Environmental benefits are calculated based on research done by the Environmental Defense Fund and
other members of the Paper Task Force who study the environmental impacts of the paper industry.

For a full list of NSP's titles, please call **1-800-567-6772** *or check out our website at:*

www.newsociety.com

NEW SOCIETY PUBLISHERS